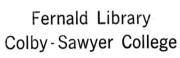

GOING METRIC

Also by **S t e w a r t M. B r o o k s**:

McBurney's Point: The Story of Appendicitis
The World of Viruses
The Cancer Story
The V. D. Story
The Sea Inside Us
Civil War Medicine
Our Murdered Presidents: The Medical Story
Basic Biology
Basic Facts of Body and Water and Ions
Integrated Basic Science
Basic Science and the Human Body
Ptomaine: The Story of Food Poisoning

GOING
METRIC

Stewart M. Brooks

SOUTH BRUNSWICK AND NEW YORK: A. S. BARNES AND COMPANY
LONDON: THOMAS YOSELOFF LTD

QC
95
B76

A. S. Barnes and Co., Inc.
Cranbury, New Jersey 08512

Thomas Yoseloff Ltd
Magdalen House
136-148 Tooley Street
London SE1 2TT, England

Library of Congress Cataloging in Publication Data

Brooks, Stewart M.
    Going metric.

    Bibliography: p.
    Includes index.
    1. International system of units. I. Title.
QC95.B76   389'.152   75-29691
ISBN 0-498-01702-8

PRINTED IN THE UNITED STATES OF AMERICA

This book is for

HELEN *and* BILL ELLIOTT

Aunt and Uncle and Very

Good Friends

CONTENTS

     Six Significant Figures   96
17   Temperature Conversion "by Fraction"   98
18   Temperature Conversion "by Decimal"   99
19   Comparison of Calculations — Customary vs. Metric 100

     Glossary   103
     Bibliography   115
     Index   119

PREFACE

In theory, anyway, the United States went metric over a hundred years ago, and by an act (1866) of Congress, no less. Indeed, the metric system is the only system that has ever received specific legislative sanction, and none other than the highly revered *yard* and *pound* are actually defined in terms of meter and kilogram. In practice many things have been metric for years, and all sorts of other things are going metric. The famed report of the U.S. Metric Study (1971), conducted by the National Bureau of Standards, concludes that "the United States should change to the metric system through a coordinated national program." And central to the success of such a program is that "early priority be given to educating every American schoolchild and the public at large to think in metric terms." The latter is the pronouncement of the U.S. Department of Commerce and an admonition underscored by the British in their march toward metrication.

The work at hand is an attempt to be "educational" in a framework that is pleasingly informative. Stated otherwise, the author looks upon measurement as a fascinating facet of civilized society and wishes to share his enthusiasm. Clearly, measurement is not a conglomeration of boring tables, but rather a story extending from the sundial to the atomic clock. And perhaps, above all, there is the element of surprise, for as it turns out the metric system is not really the metric system at all. Rather, it is the *Système International d'Unités* (International System of Units) with the official abbreviation *SI*. A few years hence, by way of example, we will be "counting *joules*," not calories, and reading the pressure gauge in *kilopascals,* not pounds per square inch. This is SI language, not metric. Finally, in a real try at being as helpful as possible, the author has employed a compartmentalized format. Part One deals with the historical development of measurement, Part Two with SI base units and their common derivatives, and Part Three with the nitty gritty; and the glossary serves as an overview, summary, and refresher.

I am deeply appreciative of the help given me by the National Bureau of Standards. The people there make you want to pay your taxes. I am particularly grateful to Louis E. Barbrow, Coordinator of Metric Activities; W.R. Telley, Chief, Office of Technical Publications; Dr. James A. Barnes, Chief, Time and Frequency Division; Kent T. Higgins, Program Information Office; and Rolfe MacCullough. I also wish to thank the American National Standards Institute and the American Society for Testing and Materials for their permission to use certain illustrations. And, of course, I thank Natalie Ann Brooks, my wife, for putting up with milligrams at breakfast, kilometers at lunch, joules at cocktails, and candelas at dinner.

INTRODUCTION

"Weights and measures," wrote John Quincy Adams in 1821, "may be ranked among the necessaries of life to every individual of human society . . . The knowledge of them, as in established use, is among the first elements of education, and is often learned by those who learn nothing else, not even to read and write. This knowledge is riveted in the memory by the habitual application of it to the employments of men throughout life."

This may be Fourth of July oratory, but just a moment's thought underscores its naked and absolute truth. From the time we rise until we climb into bed our day becomes a veritable metrology. Starting with the alarm going off at eight o'clock, the number and variety of itemizable measures is limited only by the imagination. And don't forget the *internal* environment either — pulse 72, blood pressure 120/80, red cell count 4.5 million, hemoglobin 14.5 grams, cholesterol 200 "mg%", vital capacity 3.7 liters, and on and on. And just think, you're still in bed listening to the weatherman on the radio (at 80 kilohertzes) talking about degrees Celsius and the wind velocity.

As indicated, measurement goes well beyond the popular notion of "weights and measures" — that is, a pound of butter and a quart of milk. In the real world we must measure just about everything and in a manner *everybody* agrees

11

upon. Now, after 5,000 years, this objective is almost in full sight. Though we commonly call this acme the "metric system," the truth of the matter is that the metric system proper amounts to a butter-and-milk situation, whereas the actual system — *Système International d'Unités* (SI) — provides a complete coherent system of units applicable to all physical measurement. SI covers everything, and yet in the framework of merely *seven base units,* six of which are defined in terms of a natural standard — a standard that does not and cannot change. From the base units (and two supplementary units) are *derived* four dozen or so other units for specific measurement purposes. For example, the meter and the second are base units from which we derive the "meter per second" (SI unit for velocity) and "meter per second per second" (SI unit for acceleration). What is more, certain SI units serve more than one purpose. The joule, for example, is at one and the same time a unit for work, energy, and quantity of heat. The full beauty of SI comes into focus when we consider the alternatives — the table upon table of medieval measurements of *unrelated* and *incoherent* units, often two or more being used to measure and express the same thing. For instance, there is *one* joule, but *nine* different calories! In sum, SI provides a logical and interconnected framework for all measurements in science, industry, and commerce. An SI America "is a decision whose time has come."*

In learning a new language, the brain cells are such that what you say in the new tongue involves thinking in the old tongue. And thinking in the old tongue means, ipso facto, that you understand the old tongue. Clearly this applies to "speaking SI," but with this reservation: Most people are not competent in the old tongue; indeed, they are even unaware of its official name, the "U.S. Customary System." Realisti-

* From the title of the 1971 Report on the U.S. Metric Study, *A Metric America: A Decision Whose Time Has Come.*

cally, then, we should take a look at the old tongue in regard to both content and where it came from. And this is the precise purpose of Part One. This part blends in logically, let us hope, with Part Two (SI) to afford a meaningful and enjoyable overview of "Going Metric."

GOING METRIC

Part One: H I S T O R Y

1

MINAS

Crude measurements of length, weight, and capacity probably go back to prehistoric times, and at the very early date of 3000 B.C. the great river valleys of the Tigris-Euphrates in Mesopotamia and the Nile in Egypt gave birth to civilization and certain elements of metrological sophistication. The Sumerians (of the lower Tigris-Euphrates) were the real creators of civilized society, archeologists believe, and fascinating survivals of their sexagesimal number system include modern-day angular measurement and the 360-degree circle.

Without doubt the first units of measurement were linear and related to "man as the measure." By far and away the most important was the *cubit*, an actual Egyptian hieroglyphic equal to the length of the forearm from the tip of the middle finger to the elbow (or some 17 to 22 inches). A statue of the great Sumerian king, Gudea (today in the Louvre), sports a rule on its lap representing the half cubit, such a rule and others like it indicating a strong attempt at standardization. Careful and extensive measurements of the

19

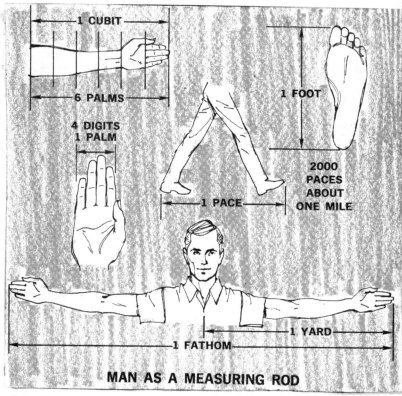

MAN AS A MEASURING ROD

The first units of measurement were undoubtedly linear and related to "man as the measure." (Courtesy National Bureau of Standard.)

Great Pyramid at Khufu show the base equal to 440 Egyptian cubits (755 feet) with a mean error in the length of the sides of only 1 part of 4,000. Other anatomical considerations include the *digit, palm, great span, little span, fathom,* and, of course, *foot.* The digit was the width of the finger; the

palm the width or length of the hand; the great span the
maximum spread between the thumb and little finger; the
little span the spread between the outstretched forefinger
and thumb; and the *fathom* the distance between the middle
finger tips of the outstretched arms (now a unit of length
equal to six feet). The foot came into full use in ancient
Greece and Rome, and both were very close to the present-
day version. According to one reliable source, the Greek
foot was 12.5 inches and the Roman foot (*pes*) 11.6 inches,
the latter being subdivided into twelve parts called *unciae*
(forerunner of the inch). Other Roman linear measures of
note were the *ingerum* (28,800 square Roman feet) and most
especially the Roman mile (*mille passus,* "1,000 paces").

Weight measures clearly came after linear measures, but
they still are ancient. And it is highly significant that for
thousands of years weighing was largely confined to precious
metals and stones, which is easy to appreciate when we consider the modern practice of buying a peck of potatoes or a
dozen oranges. Above all, this close association led to rather
extraordinary precision and the linking of weight to monetary values, a marriage epitomized by the "avoirdupois
pound" and "pound sterling." The first coins, introduced by
the Lydians in the seventh century *B.C.,* were actually
nothing more than pieces of precious metal stamped with
some sort of mark to indicate their weight. And the accuracy
was phenomenal, some ancient coins (of the same species)
varying no more than 0.00001 ounce!

The earliest basic unit of weight (and money) was the
mina of Mesopotamia. Further, the mina was subdivided
into *shekels,* and so many minas constituted a larger unit
called the *talent.* The code of Hammurabi (circa 1700 *B.C.*)
makes frequent reference to these units relative to prices
and fines. Minas, shekels, and talents together are generally
considered to be the "great parent system" for the weight

and monetary measures adopted throughout the ancient Near East. In time they evolved into the Greek *drachma* (0.01 mina), and finally the Roman *libra* (0.5 mina). The libra, like the foot, was divided into 12 *unciae,* uncia in Latin meaning "twelfth part." In the instance of the foot, the uncia evolved, as already noted, into the *inch,* and in the instance of the libra the term evolved into the *ounce.* Libra accounts for the abbreviation "lb." for pound, and well it should because it was equal to the present-day *apothecary* pound (12 ounces as opposed to the *avoirdupois* pound of 16 ounces).

The ancient measures of volume or capacity were poorly developed and poorly understood (historically) prior to the Roman Empire. The Roman unit of liquid measure was the *sextarius*, which was equal to about a pint. Six sextarii made a *congius,** and eight congii made an *amphora,* the latter being a two-handled earthenware vessel used to carry wine or oil. A great many amphorae have been recovered in modern times from sunken ships in Mediterranean waters. (Incidentally, the amphora was also employed to express the capacity of ships.) For *dry* measure the Romans used the *modius,* a "corn measure," equal to about a peck.

A crowning metrological glory of ancient Rome was the establishment of what could very well be labeled a "Bureau of Standards" at the Temple of Jupiter. In particular there was (and is) a bronze congius (dated *A.D.* 75) with the inscription saying that its "water-weight" equals 10 librae. With the fall of Rome, an extended process which began in the third century A.D., such developments passed into limbo up until about the middle of the Middle Ages.

---

* A term still used in pharmacy to signify *gallon.*

2

MINIMS

The measurement system commonly used in the United States today was basically developed during the medieval period from the intermingling of untold metrological elements. Romans, Gauls, Saxons, Britons, Arabs, and barbarians all made their mark. Considering the fact that we are talking about a thousand-year interval, the historical details are extraordinarily complex, as were the "systems" themselves. In the main this complexity and confusion arose from the intense localism which prevailed and from the tendency to measure things in units unique to a particular object or craft, a feature which still haunts us today — a peck of potatoes, a jigger of rum, a ton of coal, a yard of cloth, a cord of wood, an acre of land, and so on. The efforts of Charlemagne notwithstanding, there was little progress up until the thirteenth century, at which time most European rulers began to establish and enforce some sort of standard

23

in their particular realm of influence. No less a document than the Magna Charta (1215) refers to weights and measures. Indeed, through royal edicts, England by the eighteenth

The U.S. Customary gallon is the original "Queen Anne" wine gallon (above) which the British replaced in 1824 with the Imperial gallon. The latter measure is 1.2 times larger. (Courtesy National Bureau of Standards.)

century had achieved a greater degree of standardization than the continental countries. Further, English units were well suited to commerce and trade because they had been developed and refined to meet commercial needs. Above all, through colonization and dominance of world commerce during the seventeenth, eighteenth and nineteenth centuries, the English system of weights and measures was spread to and established in many parts of the world, including the American colonies.

The "English system" underwent certain changes along the way before becoming the British Imperial System and U.S. Customary System of modern times.* The fundamental

* And there were a number of proposed changes that never made it, including Thomas Jefferson's brain child (see illustration).

units of both systems are the *yard* and the *pound*. Further, by an agreement made in 1959 the yard and pound in both systems are defined in terms of metric standards, a rather interesting revelation when we consider that everyone is talking about "going metric." There are, however, significant differences between the U.S. and British systems. Most notably, in the British Imperial System the units of dry measure (capacity) are the same as those of liquid measure, whereas in the U.S. Customary System they are not. Further, the *troy* (or apothecary) pound was abolished in England back in 1879. Actually, it seems a little odd to be talking this way, comparing the two systems, in light of the fact that England has "gone metric." On the other hand, proclamations are not going to do away overnight with entrenched customs.

The U.S. Customary units of length of everyday interest are the *inch, foot* (12 inches), *yard* (3 feet), *rod* (5.5 yards), *furlong* (220 yards), and land, or statute, *mile* (1,760 yards; 5,280 feet).* The Roman mile, we recall, was 5,000 feet, and its conversion to the modern figure involves the early Tudors and Queen Elizabeth I. In brief, the Tudor rulers established the furlong (originally the length of the furrow made on a square field of ten acres) as 220 yards. This led Good Queen Bess to declare in the sixteenth century that henceforth the traditional Roman mile be replaced by one of 5,280 feet, thereby making the new mile *exactly* 8 furlongs.

The U.S. Customary units for area and volume (capacity) are derived from the linear by multiplication. These include the square inch, square foot, square yard, and square mile; and cubic inch, cubic foot, and cubic yard. The acre, one of the many signal oddities of the English system, is equal to 4,840 square yards or 43,560 square feet. And in regard to "liquid measure" and "dry measure" we encounter all sorts

---

* The nautical, or international mile, equals 1.1 land miles.

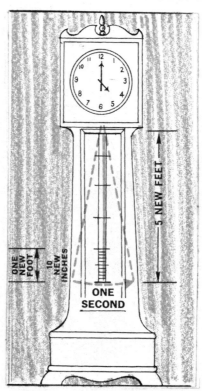

Thomas Jefferson proposed the length of a rod swinging as a pendulum with a period of two seconds, as the fundamental unit of length. This length (about 58.7 regular inches) was to be divided into 5 *new* feet of 10 *new* inches. (Courtesy National Bureau of Standards.)

of medieval antiques, especially the largely unfamiliar *minim* and *dram.* For liquid measure 60 *minims* equals one *fluid dram;* 8 fluid drams equals one *fluid ounce;* 16 fluid ounces equals one *pint;* 2 pints equals one *quart;* 4 quarts equals one *gallon;* 31 or 42 gallons ("it all depends!") equals one *barrel;* and, to add insult to injury, 63 gallons equals one *hogshead.* For dry measure, we have: Two *pints* equals one *quart;* 8 quarts equals one *peck;* and 4 pecks equals one *bushel.* (The dry pint is 33.6 cubic inches vs. 28.875 cubic inches for the liquid version.)

In the realm of weight measure, the U.S. Customary unit is the *avoirdupois pound.* Starting at the bottom, "the system" runs: 27.344 *grains* equals one *dram;* 16 drams equals one *ounce*; 16 ounces equals one *pound* (7,000 grains); 2,000 pounds equals one *short ton;* and 2,240 pounds equals one *long ton.* And then, of course, there is *apothecary,* or troy, weight, which history tells us originated at a fair in Troyes, France. This was "the system" the doctor of yesteryear used in writing his magical prescriptions. The basic unit, the grain, is the same as the avoirdupois grain, but the ounce is *heavier* and the pound *lighter* than avoirdupois! Too, we encounter a true piece of pharmaceutics — the *scruple.* The system runs: 20 *grains* equals one *scruple;* 3 scruples equals one *dram;* 8 drams equals one *ounce;* and 12 ounces equals one *pound* (5,760 grains). Interestingly, the apothecary dram and the fluid dram are *numerically* equal (60 grains vs. 60 minims) and the same goes for the ounce (480 grains vs. 480 minims) — hence, the custom of specifying "fluid dram" and "fluid ounce."

And last but by no means least we have the ancient and honorable *carat,* a unit (unto itself) of weight for precious stones. Derived from the Arabic qīrāt ("little weight"), the carat, historians tell us, relates to the carob, an evergreen tree (*Ceratonia siliqua*) with edible hornlike pods (St. John's-bread) containing numerous hard seeds embedded in pulp. Apparently the seed was the original unit, which over the centuries became first 3 1/3 grains, then 3 1/5 grains, and now, by international agreement, 200 milligrams (approximately 3 1/3 grains). So here we have the unique situation of a medieval unit "going metric" but still keeping its name.

3

MILLIMETER

Along with the explosive scientific advancements of the seventeenth and eighteenth centuries came the pressing need for a single worldwide coordinated measurement system. In 1670 Gabriel Mouton, Vicar of St. Paul in Lyons, proposed a comprehensive decimal system based on the length of one minute of an arc of a great circle of the earth; and in 1671 Jean Picard, a French astronomer, proposed the length of a pendulum beating seconds as the unit of length. (Such a pendulum would have been fairly easily reproducible, thus facilitating the widespread distribution of uniform standards.) Other proposals were made, but more than a century elapsed before any action was taken.

The real spark for the abolition of the foot and pound and all the other medieval metrologic concoctions was provided by the French Revolution, in the midst of which (1790) the National Assembly requested the French Academy of Sciences

28

to "deduce an invariable standard for all the measures and all the weights." In response, the Academy created a system that was, at once, simple and scientific. As proposed by the now long departed Gabriel Mouton, the unit of length was to be a portion of the earth's circumference. Measures for volume

The meter as proposed by the founding fathers of the metric system. (Courtesy American National Standards Institute, Inc.)

and mass (weight) were to be derived from the unit of length, thus relating the basic units of the system to each other and to nature. That is to say, the system was to be coherent. And equally critical, the smaller and larger versions of each unit were to be created by dividing or multiplying the basic units by 10 and its multiples. Thus, Mouton's decimal idea also came true. The Academy gave the name *meter* (*metre* in French) to the unit of length (from the Greek, *metron,* "a measure") and a platinum bar was to be constructed to equal one ten-millionth of the distance from the North Pole to the equator along the meridian of the earth appropriately passing through Paris. The standard unit of mass or weight was to be the *gram,* the weight of one millionth of a cubic meter of water at its maximum density, and the standard unit of liquid capacity, the *liter* (*litre* in French), was to be one thousandth of a cubic meter. Fractions and multiples of these three base units — the so-called "derived" units — were to be indicated by the appropriate prefixes; for example: millimeter (0.001 meter), centimeter (0.01 meter), decimeter (0.1 meter), dekameter (10 meters), hectometers (100 meters), and kilometers (1,000 meters).

The French National Assembly not only promptly approved the plan — now referred to as the metric system — but also "went decimal" in various other areas and directions. The monetary unit, the livre or pound, was changed to the franc (divided decimally into 100 centimes), and, lo and behold, the calendar and hours in the day were decimalized. (The ten-hour clock is a real collector's item.) Both the calendar and clock proved to be abortive and for some time it looked as though the metric system itself was headed for extinction, the National Assembly's enthusiasm notwithstanding. Actually, this revolutionary system was not on solid ground until 1840, at which time all weights and measures in France other than metric were "forbidden under penalties." In the

The French Revolution went decimal in more ways than one, as underscored above by the 10-hour watch and decimal calendar. (Courtesy National Bureau of Standards.)

United States the importance of the regulation of weights and measures was recognized in Article 1, Section 8, when the Constitution was written in 1787, but still the metric system was not *legalized* until 1866. Interestingly, it is the only system that has ever received specific legislative sanction by Congress.

At the international level the biggest step forward was the "Treaty of the Meter," concluded in Paris on May 20, 1875. Signed by seventeen nations, including the United States (Great Britain signed in 1884), the treaty reformulated the metric system and refined the accuracy of its standards; provided for the construction of new measurement standards and distribution of accurate copies to participating countries;

and established permanent machinery for the further international action on weights and measures. The new measurement standards, including meter bars and kilogram weights, were finished in 1889, at which time the United States received its copies. Four years later the Secretary of the Treasury declared the new metric standards to be the "fundamental standards" of length and mass. Thus, in 1893 the United States became an *officially* metric nation. In other words, the yard, the pound, and the other Customary units were henceforth defined as fractions of the standard metric units.

By 1900 a total of thirty-five nations had officially accepted the metric system, and today, with the exception of the United States and a few small countries, most of the world is using the system or is committed to it. In 1971 the Secretary of Commerce, in transmitting to Congress the results of a three-year study authorized by the Metric Study Act of 1968, recommended that the United States change to predominant use of the metric system through a coordinated national program. A federal decision on adopting the system is expected in the near future, but regardless of government action or inaction American industry is already on the way to metric measure. In sum, the United States is going metric and legislation will merely bring some coordination and guidance to a *fait accompli.*

The top authority on weights and measures is the International Organization of Weights and Measures founded by the Treaty of the Meter. This body consists of a General Conference of Weights and Measures, an International Committee of Weights and Measures, and the International Bureau of Weights and Measures (and its consultive committees). Headquarters is at Sèvres,* France, on a tract of land donated

* Southwestern suburb of Paris.

by the French government. Thirty-six nations are now members, including the United States. The organization recognizes only the metric system and it has defined the values of units to be used. The International Bureau of Weights and Measures, as indicated, preserves the metric standards, compares national standards with them, and conducts research to establish new standards. The National Bureau of Standards represents the United States in these activities. The General Conference of Weights and Measures, the diplomatic arm of the organization, meets periodically to ratify improvements in the system and the standards. In 1960, the Eleventh General Conference adopted an extensive revision and simplification. The name *Le Système International d'Unités* (International System of Units), with the official abbreviation SI, was adopted for this modernized metric system. Further improvements in and addition to SI were made by the General Conferences in 1964, 1968, and 1971.

SI provides a logical and interconnected framework for all measurements in science, industry, and commerce. It is built upon a foundation of seven base units — the *meter, kilogram, second, ampere, kelvin, mole, candela,* — and two supplementary units — *radian* and *steradian.* All other SI units are derived from these units. Multiples and submultiples are employed for each unit (base or derived) and named by the appropriate prefixes, ranging from 10^{-18} (*atto-*) to 10^{12} (*tera-*). For example a micrometer* is 0.000001 (10^{-6}) meter and a megameter 1,000,000 meters (10^6) meters. In the chapters to follow we shall look into the base and supplementary units and their more important and common derivatives.

* Formerly "micron."

MULTIPLES AND PREFIXES
These Prefixes May Be Applied To All SI Units

Multiples and Submultiples		Prefixes	Symbols
1 000 000 000 000=	10^{12}	tera (tĕr'á)	T
1 000 000 000=	10^9	giga (jĭ'gá)	G
1 000 000=	10^6	mega (mĕg'á)	M
1 000=	10^3	kilo (kĭl'ō)	k
100=	10^2	hecto (hĕk'tō)	h
10=	10^1	deka (dĕk'á)	da
Base Unit 1=	10^0		
0.1=	10^{-1}	deci (dĕs'ĭ)	d
0.01=	10^{-2}	centi (sĕn'tĭ)	c
0.001=	10^{-3}	milli (mĭl'ĭ)	m
0.000 001=	10^{-6}	micro (mī'krō)	μ
0.000 000 001=	10^{-9}	nano (năn'ō)	n
0.000 000 000 001=	10^{-12}	pico (pē'kō)	p
0.000 000 000 000 001=	10^{-15}	femto (fĕm'tō)	f
0.000 000 000 000 000 001=	10^{-18}	atto (ăt'tō)	a

By way of illustration: 0.001 *meter* equals a *milli*meter (symbol *mm*); 0.001 *gram* equals a *milli*gram (*mg*); 1,000 meters equals a *kilo*meter (*km*); and 1,000 grams equals a *kilo*gram (*kg*). (Courtesy National Bureau of Standards.)

Part Two: **S I**

4

SECOND

Man may never know the true nature of time, whether it is continuous or atomistic on the one hand or real or unreal on the other. According to the theory of relativity time is not an absolute or independent entity but a *variable* fourth dimension of the universe. Relativity relationships tell us that a clock would run 0.0000005 percent slower on the sun than it does on earth, and it is conceivable that our watches run very slightly faster at the top of a mountain than at the bottom of a deep mine. But the thing about time that seems to matter most is paradoxically the thing that we can do most about — measure it!

In the usual view (including that of the American Heritage Dictionary) time involves an "irreversible succession of events," the hourglass being its quintessential embodiment. Here the sand particles, the *events,* pass from top to bottom, *in succession* and, unless we "make the bottom the top,"

37

irreversibly. Thus, like all clocks, the hourglass is a well-defined event generator. The original "generator," needless to say, is the rotating earth and the apparent successive passages of the sun across a local meridian, the interval or period between passages constituting the solar day. The so-called mean solar day is the average, in length, of all solar days in a given year. It is based, interestingly, upon a hypo-thetical sun, the mean sun, defined as moving at a uniform rate along the celestial orbit in the same period as the appar-ent sun. Prior to 1956, the *second,* the base unit of time, was defined as the fraction 1/86,400 of a mean solar day. From 1956 to 1967 it was the ephemeris second, defined as the fraction 1/31,556,925.9747 of the tropical year at 00 hours 00 minutes 00 seconds December 31, 1899.* Since 1967 the second has been SI and defined in terms of an atomic clock, the acme of man's timepieces.

The first atomic clock, the ammonia clock, was invented at the National Bureau of Standards in 1948. (The "ammonia" of household fame is an aqueous solution of ammonia gas, and it is the gas we are talking about here.) Ammonia has the chemical formula NH_3, meaning that each molecule of the gas is composed of an atom of nitrogen (N) bonded to three atoms of hydrogen (H). However, these bonds are by no means rigid and the actual situation amounts to an oscillating system par excellence; that is to say, the molecule may be regarded as having its three hydrogen atoms occupying posi-tions corresponding to the three lower corners of a pyramid, with the nitrogen atom at the apex. The period process in question is the vibration of the nitrogen atom up and down through the base of the pyramid, so that its limiting lower position is at the apex of another pyramid below the first

* Tropical year is the time interval between two successive passages of the sun through the vernal equinox; also called the calendar year (365.2422 mean solar days.)

one and symmetrical with it. What is more, it vibrates at a precise frequency — 23,870,000,000 periods (cycles) per second! We know this because if radiowaves of *this* frequency are passed through a chamber filled with ammonia they are *completely absorbed.* Now, if the generator of these radiowaves (typically a quartz-crystal oscillator) is tied in with the ammonia chamber in a feedback fashion, the oscillator will be made to stay at the frequency standard of the ammonia molecule; that is, any deviation from the frequency standard will result in the incomplete absorption of radiowaves by the ammonia molecules and the concomitant passage of energy through the chamber — energy which, when electronically fed back to the oscillator, causes it, the oscillator, to correct itself. In sum, the oscillator *stays* at the frequency standard and via the appropriate gadgetry causes a second hand to move exactly *one second* during the course of 23,870,000,000 periods or cycles of radiation.

The ammonia clock gave far more accurate time keeping than had ever been known before. Some models vary no more than 3 parts per billion, but this was still not up to the demands of modern science. The ultimate in time measurement, for the present anyway, is the cesium clock, a device with a variation less than one part in ten billion! Cesium is a silver-white, very soft metal, possibly the softest of all metals, which upon exposure to air bursts into flame. Of the atoms which compose the metal, those with an atomic weight of 133 (isotope -133) vibrate, or resonate, between a "lower energy state" and an "upper energy state" at the natural and unwavering frequency of 9,192,631,770 periods (cycles) per second. In order to "observe" this atomic resonance, however, and turn it into a clock, we must employ a state selector, a microwave resonator, a quartz-crystal oscillator, a detector and a feedback circuit. The atoms, in the form of a gas (obtained by heating cesium -133 in an electric

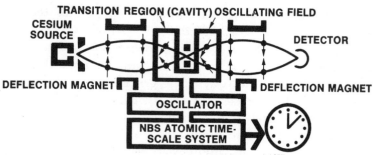

A diagram of the apparatus used to obtain the standard *second*. A frequency reading of 9,192,631,770 on the oscillator means that the duration of this many cycles is *one* second. (Courtesy National Bureau of Standards.)

oven), are first passed through the state selector which separates the lower and upper states by means of a magnetic field. Following this selection the two beams are passed through the resonator where they are acted upon by microwaves (ultrashort radiowaves) from the quartz-crystal oscillator, the atoms of the lower state making a maximal transfer to the upper state and the atoms of the upper state making a maximal transfer to the lower state when the oscillator frequency is *exactly* at atomic resonance. These transfers are monitored by the atomic detector which sends the appropriate signals to the oscillator via the feedback circuit. Whenever the oscillator starts to drift from atomic resonance (the frequency standard) the change of state is *not* maximal and the detector signal causes an increase or decrease in frequency, depending upon the direction of the drift. The upshot is a constant oscillator frequency of 9,192,631,770 periods (cycles) per second. This frequency standard was adopted by the Thirteenth General Conference of Weights

Physicist David Glaze stands by the awesome "atomic clock" (NBS-5) at the National Bureau of Standard's Time and Frequency Division (Boulder, Colorado). Without adjustment of any kind NBS-5 could run a million years and still be accurate to better than four seconds. (Courtesy National Bureau of Standards.)

and Measures (1967) and the official definition reads: "The second (symbol *s*) is the duration of 9,192,631,770 periods of the radiation corresponding to the transition between the two hyperfine levels of the ground state of the cesium -133 atom." (The two hyperfine levels are the technical expressions for "lower" and "upper" states.)

As indicated, the concept of frequency is of great significance, the frequency of a periodic phenomenon being the number of cycles (of the phenomenon) per second. The SI unit of frequency is the *hertz* (symbol *Hz*) named in honor of Heinrich Rudolf Hertz (1857-1894), the German physicist. By definition, one hertz equals one cycle per second. Thus, the radio operates on an alternating current of 60 hertzes and its dial reads in *kilohertzes* and *megahertzes.* Other derived units of the second will be discussed in later chapters.

5

METER

Light is electromagnetic radiation perceived by the unaided, normal eye. For convenience it may be thought of as waves which travel to the eye much like those of the ocean travel to the shore. Further, we speak of wavelength, or the distance from crest to crest. And central to our interest here, electromagnetic wavelength determines color. More particularly, the visible electromagnetic spectrum ranges from extreme violet, with the shortest wavelength, to extreme red, the longest. The wavelength of a monochromatic, or pure, color, therefore, could very well serve as an immutable standard of length, and such was indeed the feeling of the Eleventh General Conference of Weights and Measures held in 1960. At that time the meter was redefined as the length equal to 1,650,763.73 wavelengths (in a vacuum) of the orange-red radiation emitted by krypton 86.* This many wavelengths — accurate to one part in one billion! — equals 39.37 inches,

---

* A rare gaseous element (occurring in the atmosphere to the extent of about one part per million) used in gas discharge lamps, fluorescent lamps, and electronic flash tubes. Krypton 86 refers to a particular species (isotope) of the element which emits a *pure* orange-red light when electrified.

⁸⁶Kr ATOM

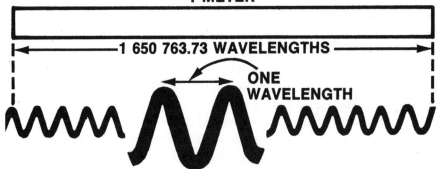

The *meter*, the SI unit of length, is the length equal to 1,650,763.73 wavelengths in a vacuum of the orange-red radiation of krypton (Kr) 86. This is equivalent to 39.37 inches. (Courtesy National Bureau of Standards.)

The light coming through the portal in the above apparatus is the orange-red radiation of krypton 86. An instrument called an interferometer (not shown) is used to measure the wavelength. (Courtesy National Bureau of Standards.)

the length of the "standard" meter bar housed at the International Bureau of Weights and Measures.

The SI symbol for meter (common international spelling, *metre*) is *m;* multiples and submultiples are indicated by the appropriate SI prefixes, ranging from *alto*meter (*am*;10^{-18} meter) to *tera*meter (*Tm*;10^{12} meters). Whenever possible SI prefixes should be used to indicate orders of magnitude, thus eliminating insignificant digits and decimals, and providing a convenient substitute for writing powers of 10. For example, 12,300 m (or 12.3 x 10^3) becomes 12.3 km.

The man in the street will find metric length to be disarmingly simple, seldom encountering anything other than millimeters, centimeters, meters, and kilometers, in that order of increasing magnitude. The decimeter, dekameter and hectometer are seldom used and in the opinion of the American Society for Testing and Materials "should be avoided where possible." Useful conversions to keep in mind are: one inch equals 2.54 centimeters; one yard equals 0.9 meter; and one mile equals 1.6 kilometers. Or, turned around, one centimeter equals 0.4 inch; one meter equals 1.1 yards; and one kilometer equals 0.6 mile.

From length we *derive* area and volume. For area we multiply *two* dimensions and label the result "square"; for volume we multiply *three* dimensions and label the result "cubic." For example, 2 in x 3 in equals 6 square inches (6 sq in or 6 in^2), and 2 in x   3 in x 4 in equals 24 cubic inches (24 cu in or 24in^3). It is certainly understood, of course, that what we multiply must be of the *same kind* — "inches times inches" and "feet times feet." By the same token metric lengths can be squared and cubed. Centimeters times centimeters gives square centimeters (cm^2); centimeters times centimeters times centimeters gives cubic centimeters (cm^3); and so on. The SI unit of area is the *square meter* (m^2); that is, "a square" one meter on a side — one meter exactly.

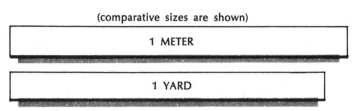

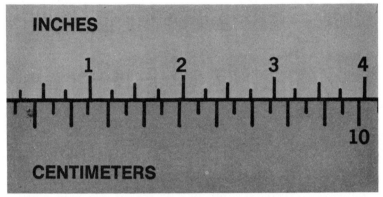

Comparative sizes of metric and U.S. Customary lengths.

The SI unit of volume is the *cubic meter* (m³); that is a cube one meter along each edge — one meter exactly.

Perhaps the great paradox of "going metric" is that we shall rarely encounter the cubic meter per se, but instead the *liter* (1) and *milliliter* (ml) — neither of which are SI units. Originally (1795) the liter (common international spelling, *litre*) was taken as a unit of volume equal to the cubic decimeter (dm³), but in 1901 the Third Conference of Weights and Measures redefined the unit as the volume occupied by the one kilogram of pure water at its maximum density under normal atmospheric pressure. The purpose of doing this was clearly to strengthen the system's "coherence." Careful determinations, however, subsequently established the liter so defined as being equivalent to 1.000028 dm^3 instead of "1dm^3, *exactly,*" as was intended. To correct the matter the General Conference in 1964 withdrew the definition and declared the word "liter" was a *special name* for the cubic decimeter or, in terms of the *official* unit, 0.001 m^3. The policy of the American Society for Testing and Materials (adopted in 1971) is that the "use of [the liter] is permitted in SI, but is discouraged, since it creates two units for the same quantity and its use in precision measurements might conflict with measurements recorded under the old definition." The position of the National Bureau of Standards is succinctly expressed in "The Modernized Metric System" wall chart (Special Publication 304 and revised October 1972), as follows: "The liter (0.001 cubic meter), although not an SI unit, is commonly used to measure *fluid* volume." (My italics.) At the practical level this is an exquisite understatement (albeit understandably so) because ounces and pints and quarts and gallons, whether it be vinegar or cough syrup or gasoline, are all "going milliliters and liters." The prescription counter went this way years ago and the package store across the street is now in the process.

For practical purposes, therefore, the liter may be taken as an "official" unit of fluid volume. It is equal to 1,000 milliliters exactly, and to 1.06 (1.056) liquid quarts. Or, stated otherwise, the quart is equal to 0.95 liter. With just a little arithmetic patience, we can show that one fluid ounce equals, in *round* figures, 30 ml, a very useful figure to keep in mind. That is, *one quart* equals 0.95 liter, or 0.95 × 1000, or 950 ml — hence, *one fluid ounce* (fl. oz.) equals "950 over 32" or 30 ml. Thus one cup (8 fl. oz.) equals, *approximately,* 240 ml, one pint (2 cups) 480 ml, and so on. Also of value

1 QUART   **1 LITER**

The liter (an unofficial SI unit of volume) versus the quart. In round figures the liter equals 1.06 quarts; more exactly it equals 1.05669 quarts. (Courtesy National Bureau Standards.)

about the house is to keep in mind that a teaspoon equals one-sixth ounce and therefore "30 over 6", or 5 ml. Further, and in round figures, one tablespoon equals 3 teaspoons and therefore 15 ml. And an interesting fact to keep in mind to underscore the magnitude of the milliliter is that *one* milliliter equals 15 to 16 drops of the size delivered by the usual eyedropper.

From length we also *derive* two "dynamic" units — speed

and acceleration. Speed is the rate or a measure of the rate of motion, especially distance traveled divided by the time of travel. If we cover 100 miles in 2 hours, for example, our speed is 50 miles per hour. Acceleration, on the other hand, relates to a *change* of speed and is determined by dividing change in speed by the time it takes for the change to occur. For instance, if we *accelerate* our car from 30 miles per hour to 60 miles per hour in 5 seconds, the acceleration is (60-30) divided by 5 — or 6 miles per hour per second (6 mi/hr/sec). Acceleration, of course, can be negative. By applying the brakes and slowing the car from 60 miles per hour to 30 miles per hour in 5 seconds, we decelerate 6 miles per hour per second. The key point to remember about acceleration is that time is expressed *twice*. In SI terms speed and acceleration boil down to *meters* and *seconds*; that is, speed is expressed as meters per second (*m/s*) and acceleration as meters per second per second (meters divided by seconds squared, or m/s^2).

6

KILOGRAM

The kilogram (symbol *kg*)* is the fundamental unit of *mass,* and the only base unit still defined by an artifact. The artifact — the kilogram — is a cylinder of platinum-iridium alloy kept by the International Bureau of Weights and Measures at Sèvres, France. A duplicate — Prototype Kilogram 20 — in the custody of the National Bureau of Standards (Washington, D.C.) serves as the mass standard for the United States. The seal of Prototype 20 (along with the now "defunct" Prototype Meter 27) was broken by President Benjamin Harrison on January 2, 1890. Taken to France almost fifty years later (1937) for recomparison with the international standard, it had changed only one part in fifty million.

Although for everyday matters we equate mass and weight, they are clearly not the same. We *weigh* less on the moon and nothing at all in space; that is to say, weight is a *variable.* More particularly, weight is a variable force (W) equal to mass

---

* The kilogram (1,000 grams) is the only SI unit defined "in the plural." Originally, in the old metric system, the gram was the unit of mass.

This is prototype kilogram 20, the *mass* standard (for the United States) in the custody of the National Bureau of Standards. It is an exact duplicate of the mass standard kept by the International Bureau of Standards at Sèvres, France. (Courtesy National Bureau of Standards.)

(M) — *a constant* — times gravitational acceleration (g) (or W = Mg). Stated otherwise, weight is proportional to mass, and mass is "quantity of matter" — a quantity unchanged by its location. The value of g fluctuates by over 0.5 percent at various points on the earth's surface, and therefore the difference of local g from the agreed standard value must be taken into account for precise measurements where g is involved. In weighing on balances or scales in which a stan-

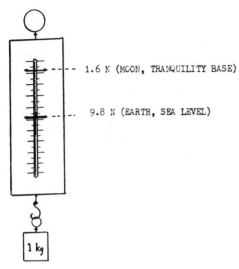

1.6 N (MOON, TRANQUILITY BASE)

9.8 N (EARTH, SEA LEVEL)

1 kg

The weight of a kilogram mass on the moon and on Earth. Since weight is a force, it is expressed in newtons (N). (Copyright American Society for Testing and Materials, reprinted by permission.)

dard mass balances the measured mass, either directly or indirectly, the effect of gravity cancels out and mass and weight become one and the same. Thus, for practical purposes, we take the kilogram to be the fundamental unit of mass *or weight*.

The kilogram is equal to 2.2046 pounds (2.2 pounds, rounded off); turned around, one pound equals approximately 0.45 kilogram (454 grams). One ounce equals 28.3495 grams (28 grams rounded off). We are, of course, referring to the U.S. Customary or *avoirdupois** pound and ounce. The *apothecary* ounce equals approximately 31.1 grams,

---

* Literally, "to have weight."

and the apothecary pound approximately 373.2 grams. In
commerce and about the house, weight metrication essen-
tially boils down to kilograms, grams, and milligrams, the
last named relating for the most part to the medicine cabinet.
The tranquilizer *Valium,* for instance, is marketed in 2-mg,
5-mg, and 10-mg tablets. And, for the sake of interest, if
you wish to convert your aspirin to milligrams, simply
multiply grains by 65. For example, the popular five-grain
tablet metricizes to 325 mg.

Closely allied to mass is *density,* or the mass per unit
volume of a substance under specified or standard conditions of pressure and temperature. The proverbial "pound
of lead" and "pound of feathers" certainly weigh the same
(under the same conditions of gravity), but our intuition
correctly informs us that the latter is decidedly *less dense*
than the former. And, as indicated, this can be reduced to
actual figures by dividing weight by volume. In our familiar
(U.S. Customary) way of handling density we divide pounds
by cubic feet and come out with pounds per cubic foot
(or lbs/ft^3). And to simplify the procedure in actual prac-

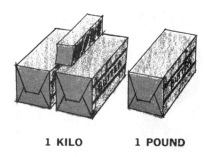

1 KILO   1 POUND

In round figures one kilogram equals 2.2 pounds. On a basis of accuracy
to six significant figures, one kilogram equals 2.20462 pounds.

tice we merely have to weigh accurately *one* cubic foot of the substance in question. For example, one cubic foot of fresh water and one cubic foot of sea water weigh, respectively, 62.4 pounds and 64 pounds, and have densities, respectively, of 62.4 lbs/ft^3 and 64/ft^3.

The SI unit of density is kilogram per cubic meter; the formula is kilograms divided by cubic meters (kg/m^3). For most purposes, however, scientific and otherwise, this unit is reduced to much smaller SI terms. Gases are usually expressed in grams per liter (g/l), and liquids and solids in grams per milliliter (g/ml). And an interesting and highly practical fact to commit to memory here is that water has a density of *one*; that is, one milliliter weighs one gram or, turned around, one gram occupies a volume of one milliliter. The density (in g/ml) of any liquid or solid compared to the density of water (1g/ml) provides us with what is called *specific gravity*. More particularly, we divide the density of the substance in question by 1g/ml and in so doing cancel out "grams per milliliter." In other words, specific gravity is a number — a number *without* units. For example, by dividing the density of mercury (13.54 g/ml) by one (1g/ml) we get 13.54; that is to say, mercury is 13.54 times heavier (more dense) than water. Clearly, substances with specific gravities *greater* than one are heavier than water, and those with specific gravities *less* than one are lighter than water. Thus, lead sinks and cork floats.

Turning now to more dynamic matters, we encounter the *newton* (symbol N), a derived unit from which, in turn, arise other fundamental SI measurements. The newton, named after Sir Isaac Newton (1642-1727), deals with force, and by force we mean a directional (vector) quantity that tends to produce an acceleration of a body in the direction of its application. *One* newton is the force which, when applied to a *one*-kilogram mass, gives the kilogram mass

The *newton* (*N*), the SI unit of force, is the force which, when applied to a 1-kilogram mass, will give the kilogram mass an acceleration of 1 meter per second per second. (Courtesy National Bureau of Standards.)

an acceleration of *one* meter per second per second. Stated in more familiar language, a force of one newton increases the speed of a one-kilogram body one meter per second (m/s) every second. To compute newtons we multiply kilograms times meters and divide by seconds squared, or $N = kg \cdot m/s^2$.

Hand in hand with force goes pressure, the latter being defined as force applied over a surface. Pressure is measured as force per unit of area and in the language of SI becomes *pascals* (*Pa*), after the French philosopher and mathematician Blaise Pascal (1623-1662). One pascal is a force equaling one newton per square meter. Thus, to compute pascals we divide newtons by square meters, or *Pa* 8 N/m^2. Vis à vis "pounds per square inch," the pascal is a small value, one pound per square inch equaling 7 kilopascals. An automobile tire, for example, requiring 30 pounds of pressure would be pumped up to 210 kilopascals.

Dividing newtons by square meters yields pascals; and now we shall see that multiplying newtons by meters (*N·m*) yields *joules* (*J*), the SI unit of *energy*. In brief *J=N·m*. One joule, named after the British physicist James Prescott Joule (1818-1889), equals the work done when the point of application of a force of *one* newton is displaced by *one* meter in the direction of the force. And by virtue of the definition, we note that the joule also becomes the SI unit of *work*. In our "English system" we compute work by multiplying pounds times feet to yield foot-pounds. By way of example, in lifting a 20-pound weight 5 feet we do 100 foot-pounds of work. Now, one foot-pound equals 1.3558 joules, so this particular effort in SI becomes 135.58 joules.

Further, the joule is the SI unit for the *quantity* of heat and will eventually replace the calorie and Btu (British thermal unit). The *large* calorie of diet fame is the amount of heat needed to raise the temperature of one kilogram of water 1°Celsius (centigrade). In terms of SI this amounts to 4,186 joules or, in preferred parlance, 4.186 kilojoules (4.186 kJ). For instance, an ounce of *Wheaties* supplies 101 calories or, in round figures, 423 kilojoules (423 kJ). The British thermal unit (*Btu*), the old stand-by in expressing the heat content of fuel, is defined as the amount of heat required to raise the temperature of one pound of water one degree Fahrenheit. In terms of SI, one Btu equals 1,055 joules or 1.055 kilojoules. For example, one pound of good-grade coal yields about 15,000 Btu or 15,825,000 joules or, in the preferred unit, 15.825 megajoules (15.825 MJ).

By dividing joules by seconds (*J/s*) we obtain *watts* (*W*), the SI unit of *power* named in honor of the Scottish engineer and inventor James Watt (1736-1819); in the singular, *one* watt equals *one* joule per second. Power is the rate at which work is done, the familiar (U.S. Customary) unit being the *horsepower* (hp). One horsepower equals 33,000 foot-pounds (of work) per minute, or 550 foot-pounds per second, the figure Watt arrived at in seeing how much work a horse could do in a minute in drawing coal from a coal pit. In SI language, one horsepower equals 745.7 watts, the 80-horsepower engine thus becoming the 59.66 kilowatt engine. The watt, of course, applies to *any kind* of power as underscored by the light bulb. The 75-watt bulb, for example, amounts to the expenditure of 75 joules of electrical energy per second; that is, 75 "electrical joules" become heat and light energy capable (were they harnessed) of doing 75 joules of work per second.

7

M O L E

Years and years ago physics was commonly referred to as "natural philosophy," and there is indeed much to recommend this appellation. The subject certainly provokes philosophic thought in a variety of situations, the mundane, upon reflecting, being no less intriguing than quanta or quasars or quarks. Other things being equal or taken into account, Nature says: No matter how long you boil water the mercury remains at $100°C$; no matter how much ice you add to ice water the mercury remains at $0°C$; no matter how hard you beat a drum the sound travels at the same speed; no matter how much you run and jump, fat accumulates if joules eaten exceed those lost, "diet revolutions" notwithstanding; and so on.

In order to really appreciate the mole, the SI base unit for *amount of substance*, we must perforce consider a bit of natural philosophy involving this value. And we can do no better than to scrutinize the freezing point of water. Chemically pure water freezes at $0°$ Celsius (or at the familiar $32°$ F). Dissolving something — anything — in water *lowers* its freezing point, which is precisely why we add antifreeze to the car radiator. The extent of this lowering, however, relates to the nature of the substance (called the solute) dissolved in the water. By way of example, consider ethyl alcohol and sugar. Ethyl alcohol has the chemical formula C_2H_5OH, which means that each and every molecule contains 2 atoms of carbon (C), 6 atoms of hydrogen (H), and 1 atom of oxygen (O); sugar (sucrose) has the chemical formula $C_{12}H_{22}O_{11}$, which means that each and every molecule contains 12 atoms of carbon, 22 atoms of hydrogen, and 11 atoms of oxygen. Further, ethyl alcohol and sugar, like all chemical compounds, have a particular *molecular* weight, which is found by adding together the *atomic* weights (obtainable from appropriate tables). Atomic weights, of course, are not *absolute* weights, but *relative* values, which means that molecular weights, too, are relative. The molecular weight of ethyl alcohol is 46, and the molecular weight of sugar is 342, meaning that a molecule of sugar is seven times heavier than a molecule of ethyl alcohol.

In the framework of the above knowledge, let us dissolve 46 grams of ethyl alcohol in 1 kilogram of water, 342 grams of sucrose in another kilogram of water, and then proceed to freeze both solutions. (Note that the weight used corresponds to the molecular weights; this is the so-called "gram molecular weight.") Careful readings will show that *both* solutions freeze at $-1.86°$ C (that is, $1.86°$ *below* $0°C$). This appears to be contrary to common sense; after all, we used much more sugar than alcohol, but facts are facts and we must come up with an explanation. Actually, the key relates to the greater amount of sugar needed (vis à vis alcohol) to produce the *same* effect; that is, since both substances are composed of molecules, the results suggest that what really counts is the *number* of molecules rather than their size or kind. And that is exactly what much experimentation and brain power has proved: a gram molecular weight of alcohol and a gram molecular weight of sugar contain the *same number* of molecules. What is more — and lo and behold — gram molecular weights of *all compounds* contain the *same number* of molecules, a number which, fantastic as it may seem, was determined back in 1908 by the French physicist Jean Baptiste Perrin (1870-1942). Traditionally (and somewhat paradoxically) referred to as Avogadro's number, in honor of the Italian physicist Amedeo Avogadro (1776-1856), the number is 6.023×10^{23} or, in "long hand," 602,300,000,000,000,000,000,000.

We now rejoin the mole (pronounced like all other "moles") and do so most naturally because in actual practice it stands for gram molecular weight. At first a nickname, it has enjoyed official status in chemistry since the turn of the century. But the official SI definition makes no reference to the gram molecular weight because "amount of substance" deals with *elements* as well as compounds. The SI mole (symbol *mol*) is the amount of substance of a system that contains as many particles as there are atoms in 0.012 kilogram of carbon 12.* Well, 0.012 kilogram equals 12 grams or the *gram atomic weight* of carbon 12, the amount of carbon which contains the Avogadro number of atoms. The gram atomic weight of *any* element contains the Avogadro number of "free particles," atoms in some cases and molecules in others. For example, a gram atomic weight of neon (Ne), an element in which the atoms occur individually, contains the Avogadro number of atoms, whereas a gram atomic weight of chlorine, an element in which the atoms occur in pairs, contains the Avogadro number of molecules. In sum, then, the SI mole is an amount of substance containing the Avogadro number of free particles. For elements

* That is, carbon whose atoms are all "isotope 12" and thus have an atomic weight of 12.

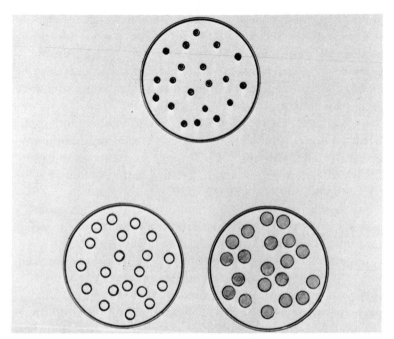

The mole, the SI unit of the amount of substance, relates to the number of particles without regard to their size or kind. Assuming for the sake of argument that 5 particles constitutes "one mole," then in the above illustration each of the three large circles (of 20 particles apiece) represents 4 moles of substance.

this amounts to the gram *atomic* weight; and for compounds, the gram *molecular* weight.

In practice, the SI mole system works this way. Suppose we have, say, 92 grams of ethyl alcohol and wish to express this in moles. We simply find the gram molecular weight and divide this into 92; or "92 over 46" gives 2 mols. Again, 513 grams of sugar (gram molecular weight 342) is "513 over 342" or 1.5 mols. And above all, let us underscore the fact that *one mole* of ethyl alcohol and *one mole* of sugar each contain the Avogadro number of molecules. For the sake of explicitness the mole should always be qualified as to the elementary entity we are talking about. As indicated, the mole may refer to atoms, molecules, ions, electrons (or other particles) or specified groups of such particles.

Hand in hand with the amount of substance is concentration, a situation analogous to heat and temperature. For example, it is one thing to dissolve an ounce of salt in a pint of water, and quite another to dissolve an ounce of salt in a quart of water. Both solutions contain the same amount of salt, but the latter is certainly *less* concentrated. The SI unit of concentration is the mole per cubic meter (mol/m^3), an expression embracing physical systems all the way from the number of molecules of sodium chloride in a cubic meter of sea water to the number of atoms of neon in a cubic meter of air. A more useful, albeit non-official, expression in the chemical laboratory is the "molar solution," which is defined as the number of moles of a compound *per liter* of solution. For example, one mole of salt (58.5 grams) per liter of solution is "1 molar" (1mol), 2 moles (117 grams) per liter 2 molar (2mol), and so on.

8

K E L V I N

Sunlight streaming through a rent in the drawn shade affords an excellent introduction to our present-day understanding of heat in general and the kelvin in particular. In just a moment we will appreciate the connection.

The dancing dust particles seen in a path of light is a classic natural phenomenon first observed in 1827 by the botanist Robert Brown (1773-1858). Brown, interestingly, noted this motion, now known to one and all as Brownian motion (or movement), during the course of the microscopic examination of pollen grains suspended in water. Dust particles and pollen grains, of course, do not dance under their own steam, but rather because of the unceasing bombardment by the invisible molecules of the liquid or gas in which they are suspended. Indeed, Albert Einstein, in 1905, developed a theory for Brownian motion on the assumption that the suspended particles behaved as though they were large-sized molecules. But for our purpose here the central point is this: The molecules of which a substance is composed, whether the substance be a solid, liquid or gas, are in *constant motion.*

Molecules are in constant motion and accordingly possess energy or, to be more precise, kinetic energy. What is more, the difference among the three states of matter is related to this motion, the molecules of a gas having the most motion — and energy — and the molecules of a solid the least motion and energy. Applying heat to ice underscores the point; that is, the molecules pick up speed to become water, and with continued heating pick up even more speed to become steam. Thus, heat becomes molecular motion and kinetic energy; or, to be blunt about it, heat *is* energy (certainly a form of energy). The proof is overwhelming in and out of the laboratory. For example, it gets warmer when it snows because of the loss of molecular motion and *energy* — heat! — in going from the liquid to the solid state. Again, a steam burn is more serious than a boiling-water burn because the molecules give up energy — heat! — in the process of condensation. Again, an ice cube is cold to hold because its molecules absorb heat from the hand in the process of becoming the more energetic molecules of water.

If heat is a form of energy, as evidenced by molecular motion, what, then, is temperature? Are they one and the same? The answer is no. Rather they are two aspects of the same phenomenon, as evidenced by a cup of *hot* tea and a bathtubful of *warm* water. The molecules in the tea are understandably in greater motion than those in the bathtub. On the other hand, there are trillions upon trillions more molecules in the bathtubful of water, with a combined total motion, total energy — total heat — understandably much greater than the total heat of the cup of tea. Thus, temperature, which the dictionary defines as the degree of hotness or coldness, relates theoretically to the degree of motion or kinetic energy of the *individual* molecules.

To measure temperature we use the thermometer, an instrument not without historical interest. It was discovered in 1600 by none other than the great Galileo (1564-1642), and a replica can be easily fashioned right at home from a small flask, a one-hole rubber stopper, and a glass tube (about 2 feet in length). The tube is fitted to the stopper, the latter in turn to the flask, and the assembly then secured upright with the free end of the tube immersed in a glass of water. A match is now struck and the flame held near enough to the flask to force out a dozen or so bubbles of air, after which (after removal of the heat) the water will rise in the tube, somewhere near the middle of the tube. From this point on the water will go up or down the tube depending on the temperature of the room. If the temperature goes up, the air in the flask expands and pushes the water down; if the temperature goes down, the air in the flask contracts and the water rises. Indeed, by adding a drop or two of ink to the water and fastening a ruler along the side of the tube we shall have made an instrument of some sophistication. Although it must have been a little awkward, physicians used Galileo's thermometer, or thermoscope as it was then called, to tell if the patient had a fever. First, they would place the glass bulb in their own armpit and then in that of the patient for comparison.

The development of the thermometer from 1600 on was unbelievably slow. It took years to arrive at mercury (as a replacement for air) and more than a century to perfect a *standard* thermometric scale. The late date was 1722 and the man, the German physicist Gabriel Fahrenheit (1686-1736). Using a mercurial thermometer, Fahrenheit marked off as zero the lowest temperature obtainable with a mixture of salt and ice, and the freezing and boiling points of water as $32°$ and $212°$, respectively, the same as on the modern version of this famous scale. Twenty years later, in 1742, the

25 DEGREES FAHRENHEIT

25 DEGREES CELSIUS

A vivid comparison of Fahrenheit and Celsius. (Courtesty National Bureau of Standards.)

Swedish astronomer Anders Celsius (1701-1744) "went metric" by taking the boiling point of water as 0° and the freezing point an even 100°, and in 1750 Stromer switched these values to give us our present-day Celsius ("centigrade") scale.

Since time immemorial, it seems, the conversion of Fahrenheit to Celsius and Celsius to Fahrenheit has been un-necessarily and unduly ritualized, and now that we are in the process of "going metric" the situation can be expected to worsen unless well meaning teachers and folksy weather-men stop making it complicated. In fact, it is easier than slipping on ice. "To go from °C to °F," we multiply by 1.8 and add 32; "to go from °F to °C," we subtract 32 and divide by 1.8. And to check our memory we use the boiling points, 100° and 212° (or the freezing points 0° and 32°). For example, does 1.8 times °C plus 32 really give me °F? Well, 1.8 × 100 equals 180, and 180 plus 32 equals 212°. Yes, it does. Again, does °F minus 32 divided by 1.8 give us °C? Well, 212 minus 32 equals 180, and 180 divided by 1.8 equals 100°. Yes, it does. And as far as minus values are con-cerned, we must use our common sense. For example, −10°C on the Fahrenheit scale equals 1.8 times (−10) plus 32, or −18 + 32, or 14°F. After all, −10°C is below freezing on the Celsius scale and *should be* below freezing on the Fahrenheit. And so it is ("18° below"). A very interesting temperature, incidentally, is −40° — it is the same on *both* scales!

Though we are now in the process of "going Celsius," the scale is by no means SI. Like the liter it is permitted, but not preferred. In the words of the American Society for Testing and Materials: "The SI temperature scale is the International Thermodynamic Temperature Scale, and the SI unit is the *kel-vin;* therefore kelvins should be used to express temperature and temperature intervals as the *preferred* unit. However, wide use is made of the degree Celsius, particularly in engin-eering and in nonscientific areas, and it is *permissible* to use the Celsius scale where considered necessary." (My italics.)

And so this brings us to a discussion of the kelvin, the SI unit of temperature named in honor of William Thompson Kelvin (1824-1907). By official definition, the kelvin (sym-bol K), is the fraction 1/273.16 of the thermodynamic temp-

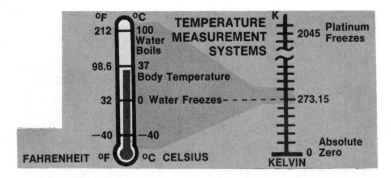

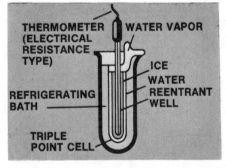

(Above) A comparison of Fahrenheit, Celsius, and Kelvin scales.

(Below) A diagram of the apparatus ("cell") used to determine the standard temperature at the triple point of water. When the cell is cooled until a mantle of ice forms around the reentrant well, the temperature at this interface of solid, liquid, and water is 273.16 K. Thermometers to be calibrated are placed in the reentrant well. (Courtesy National Bureau of Standards.)

erature of the triple point of water. The triple point of a substance refers to the coexistence of its three states, a

A photograph of the apparatus ("cell") used to determine the standard
temperature (273.16K) at the triple point of water.

natural phenomenon which occurs only at one pressure and one temperature; for water, the liquid, ice, and vapor co-exist at a pressure of 0.006 atmosphere and at a temperature of 0.01°C. On the thermodynamic scale this corresponds to 273.16°, which gives a value of *one* (exactly) when multiplied by 1/273.16 (as per definition). The thermodynamic, or absolute, scale starts at *absolute zero* (−273.15°C or −459.67°F), the temperature at which molecules possess minimal energy. Indeed, this is why the absolute scale is the preferred scale — that is, zero temperature is "zero energy" and there are *no negatives*.

In way of refinement, one kelvin (1K) means one division on the absolute scale and one division (*one degree*) on the Celsius scale. A given *temperature*, however, must take into account the value 273.15; that is, degrees Celsius (°C) equals kelvins minus 273.15 or, turned around, kelvins equals degrees Celsius plus 273.15. For example, 0°C equals 273.15K (0°+273.15); 0K equals −273.15°C (0−273.15); and so on. Note that we do not speak of or indicate "degrees kelvin"; note, too, that kelvin is not capitalized and that the symbol (*K*) is capitalized and without a period.

Turning to the *quantity* of heat, we encounter the "battle of the bulge" and the calorie. Actually, the calorie (from the Latin *calor*, heat) comes in several kinds and sizes, and there is sometimes no little confusion as to the one under discussion. The unit is certainly on the way out, but how long it will take to say "watch your joules" is anybody's guess.

Unless otherwise specified a calorie is the amount of heat required to raise the temperature of one gram of water 1°C from a standard initial temperature, especially from 14.5°C to 15.5°C. By way of example the amount of heat needed to raise the temperature of 10 grams from 14.5°C to 15.5°C is 10 calories; the amount of heat needed to raise the temperature of 10 grams of water from 14.5°C to 16.5°C is 20 calories; and so on. This, however, is not the calorie referred to on the box of cereal. Alas, the calorie of diet fame, the "large calorie" (a natural pun if there ever was one) is the amount of heat required to raise the temperature on one *kilogram* of water by 1°C at one atmosphere of pressure. One kilogram, of course, equals 1,000 grams, meaning that the large calorie (sometimes called "kilogram calorie") equals 1,000 regular or "small calories." For what it is worth, people in the know abbreviate the small calorie *cal*, and the large calorie *Cal*.

As we learned earlier, the SI unit for work and energy of any kind is the joule (J), and since heat is a form of energy the joule becomes the obvious and natural unit for the quantity of heat. Put otherwise, whereas the calorie confines the implications of heat, the joule interrelates and equates heat with work and all other forms of energy. In way of conversion, the small calorie (cal) equals 4.19 joules, and the large calorie (Cal) 4.19 kilojoules. One ounce of Wheaties, for instance, supplies 101 Cal or 423.2 kilojoules.

9

CANDELA

To say that a given light source, whether it be a kerosene lamp or klieg light, is so many *candlepower* conceivably confuses no one. After all, the candle is just about synonymous with what we call light and most certainly is as much revered at "GE" as anywhere else. Clearly, candlepower means luminous intensity or, more particularly, luminous intensity expressed in *candles*. Stated otherwise, the older we get, the more *luminous* our birthday cakes become. Candles, of course, come in all sizes, so common sense tells us to establish a *standard* candle, and this was indeed done long ago. The standard candle (symbol c) was originally defined as one made of spermaceti (the waxy substance obtained from the head of the sperm whale) and which burned at a rate of 120 grains (about 0.3 oz.) per hour. Thus, a lamp giving ten times as much light as such a candle was said to be 10 candlepower. But gradually, both here and abroad, the flame gave way to the incandescent filament ("standard lamp"), and finally in 1948 the General Conference of

Weights and Measures made the candle SI.

The SI candle is the euphonious *candela* (symbol *cd*). It is defined in the most sophisticated and precise terms known to modern physics, and here it is: The candela is the luminous intensity of 1/600,000 of a square meter of a blackbody at the temperature of freezing platinum (2045K) under a pressure of 101,325 newtons per square meter. In more down-to-earth terms, visualize a furnace with a single opening of one square centimeter, an opening which appears pitch

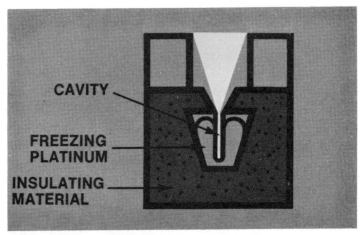

A diagram of the apparatus used to establish the *candela*, defined as the luminous intensity of 1/600,000 of a square meter of a blackbody at the temperature of freezing platinum.

black (a "blackbody") when the furnace is cold. If we now heat the furnace to 2045K (or to the more familiar 3,224°F), the blackbody becomes a luminous body — 1/60 square centimeter of which (1/600,000 meter) represents *one* candela. This luminous intensity is equal to 0.98 candle (à la spermaceti) or, turned around, one candle equals 1.02 candela.

Lest we associate "freezing" exclusively with the refrigerator, let us not forget that freezing and melting are the sides of the same physical coin and occur at the same temperature. Ice and water remain side by side at 32°F, irrespective of whether the ice is melting or the water is freezing. Platinum melts at 3,224°F and solidifies — *freezes* — at 3,224°F (1773°C; 2045K) at the specified pressure. As to why the definition speaks of freezing rather than melting relates to certain technical and practical matters. In regard to blackbody, a perfect blackbody is a flawless emitter of radiant energy — in the case at hand the emitter of visible radiant energy generated at the *immutable* temperature of freezing platinum.

Absolute photometric measurements by comparison with the actual blackbody candela can only be undertaken by a few well-equipped laboratories, the accuracy being somewhat better than 1 percent. The results of these measurements are maintained by means of incandescent lamps, and it is these lamps which in actual practice constitute the standards of luminous intensity. Some idea of how a standard lamp can be used to establish the value of a lamp of unknown candle power is provided by the so-called "oil spot" experiment. An oil spot on heavy white paper placed near a light source appears dark when viewed from the illuminated side and bright from the other side. If such a paper is placed between two sources of light, say two lamps, the action just described takes place for each of them. Further, if the paper is moved away from one of the lamps and toward the other, the spot grows brighter or darker, depending on which side is viewed. At one position, and one position only, the spot will just about disappear, and when this occurs the respective distances of the lamps to the paper will tell us the candlepower of the lamp in question; that is to say, they tell us the candlepower in the framework of this law: The intensity of light at different distances from the source varies *inversely* as the square of the distance. Thus, the candlepowers of the lamps are in the same ratio as the squares of the distances to the paper. For example, if the standard lamp is 2 candela and one foot from the paper, and the unknown lamp 2 feet from the paper, then the latter must be 8 candela.*

Hand in hand with the candelas are *lumens*. The flow of light from its source is referred to as luminous flux and the SI unit involved in its measurement is the lumen (*lm*). One lumen is the luminous flux emitted in a solid angle of one *steradian*** by a point source having a uniform intensity of one candela. A light source having an intensity of *one* candela in *all directions* provides a luminous flux of 4π (12.56) lumens. A 100-watt light bulb emits about 1700 lumens.

The illumination of a surface one foot distant from a source of one candela is referred to as a *footcandle* (or sometimes "candle-foot"), an excellent example of an intermingling between the SI and English systems. The marriage was inadvertent, of course, and occurred when spermaceti gave way to the blackbody. The SI unit of illumination is the *lux* (*lx*), which is the luminance produced by a luminous flux of *one* lumen uniformly distributed over a square surface of *one* square meter (or lm/m^2). Whereas the candle and candela are just about equal, the footcandle is ten times the lux (10.76 times, to be exact). The lux, incidentally, is another example of SI unit interplay, involving, as it does, the meter, the lumen, and indirectly (via the lumen) the candela and the steradian.

$$*\frac{2}{1^2} = \frac{X}{2^2} \quad \text{or} \quad \frac{2}{1} = \frac{X}{4}$$

** To be discussed in the next chapter.

10

AMPERE

According to the *American Heritage Dictionary*, electricity is "the class of physical phenomena arising from the existence and interaction of electric charge." Thus, "electric charge" is the crux of the matter, notwithstanding the fact that scientists do not know exactly what electricity is. Stated another way, electricity cannot be satisfactorily visualized but we can indeed produce it, control it, and measure it — and we do so in the context of what charges appear to be. At the very least, and as we shall now see, the electric charge is the acme of simplicity.

The essence of the electric charge is that it appears in *two* forms. When a glass rod is rubbed with silk and placed in a freely rotating wire stirrup, it is *repelled* by another silk-rubbed glass rod. Similarly, ebonite rods rubbed with cat fur repel one another. On the other hand, a glass rod and ebonite rod (when rubbed as described) *attract* one another. Thus, the ebonite rod is charged one way and the

77

glass rod another, and to underscore the point Benjamin Franklin labeled the former *negative* and the latter *positive*. Franklin considered electricity to be an "invisible fluid" of which a charged body either had too much or too little; accordingly, electricity flows from positive ("too much") to negative ("too little"). Why Franklin picked glass to be positive and ebonite to be negative is an interesting question because we know today that his "positive" is today's "negative" and vice versa.

In the framework of present-day knowledge, the classic glass-ebonite experiment boils down to the electron, a subatomic *negative* particle of electricity. Further, electrons are easily dislodged and removed from parent atoms, such a loss resulting in a *deficiency* of electrons. Remembering that a *neutral* body contains an *equal* number of positive and negative particles, a deficiency of the latter means an excess of the former and consequently an overall positive charge. By the same token the lost electrons always have to be gained by another body, thereby creating an excess and a negative charge. By rubbing the ebonite rod with cat fur we transfer electrons from the fur to the ebonite, making the rod negative and the fur positive. Accordingly, one ebonite rod repels another ebonite rod because they both have the *same* charge ("like charges repel"). By rubbing the glass rod with silk, we transfer electrons from the glass to the silk and thereby make the rod positive and the silk negative. Accordingly, one glass rod repels another glass rod (again, "like charges repel"); further, the ebonite rod and the glass rod attract each other because "unlike charges attract".

And now to continue. When the charged ebonite rod and the charged glass rod are connected by a copper wire a spark is produced and both rods lose their charge; that is, the excess electrons on the ebonite rod rush with incredible speed through the wire to the electron hungry glass rod and thereby re-establish neutrality. This flow of electrons is an electric current, the effect we unleash in turning the ignition key or pushing down the toaster. Clearly such currents vary in strength and intensity, as many of us discover in tampering with a battery on the one hand and a floor plug on the other. Further, an electric current passing through a wire creates a magnetic field about the wire, a field that will attract or repel another electromagnetic field (depending on the directions these fields take); further still, the greater the current the greater the magnetic field. By quantifying this magnetic effect we at one and the same time quantify and measure electric current, and this brings us to the *ampere* (or "amp," to use the vernacular). Named after the French mathematician and physicist André Marie Ampère (1775-

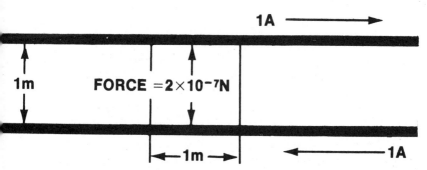

The *ampere* is defined as that current which, if maintained in each of two long parallel wires (above) separated by one meter in free space, would produce a force between the two wires (due to their magnetic fields) of 2×10^{-7} newton for each meter of length.

1836), the ampere is that constant current which, if maintained in two straight parallel conductors of infinite length, of negligible cross section, and placed one meter apart in a vacuum, would produce between these conductors a magnetic force equal to 2 x 10^{-7} (0.0000002) newton per meter of length. Note once again the tie-in with the other base units — the newton and the meter. Derived units (of the ampere) include, among several others, the *volt,* the *watt,* the *ohm* and the *siemens.*

The volt is a measure of the force — electromotive force or voltage — responsible for the flow of electrons, or electric current. Stated otherwise, it is a measure of the potential difference between a negative charge and a positive; the greater the difference between the two, the stronger the electric current when the charges are connected by a wire. For example, if negative charge *A* vis à vis positive charge *B* represents, say, a difference of a trillion electrons, then its "potential" is greater than if the difference were, say, a billion electrons. For practical purposes, the expressions "potential difference," "electromotive force," and "voltage" are synonymous, with the last named being by far the most familiar about the house — the classic "110 volts" we tap from the electric company in turning on the 60-watt bulb.

Named after Alessandro Volta (1745-1827), the Italian physicist, the volt (*V*) is equal to the difference of electrical potential between two points on a conducting wire carrying a constant current of *one* ampere when the power dissipated between the points is *one* watt; that is, volts equals watts divided by amperes, or $V=W/A$. Note the usefulness of this formula in determining "amps" ($A=W/V$) and watts ($W=VA$). The latter formula, incidentally, shows that our electric bills (in kilowatt hours) relate fundamentally to "amps," or the electric current which *we allow* to flow through the floor lamp, television set, waffle iron, and what have you. Put another way, house voltage is a "constant" (110-120V), whereas amperage is a variable and depends upon the appliance. By way of example, a 60-watt bulb and a waffle iron both operate on 110-120 volts but "draw" 0.5 amp and 12 amps, respectively; in terms of power (and cost) this amounts to 60 watts (the bulb says so!) and 1,400 watts (1.4 kilowatts).

The difference in the amps drawn by our various appliances leads very naturally to the topic of electrical *resistance* — specifically, to the resistance which electrons encounter in their flow through the appliance's circuitry. Toasters, waffle irons, electric heaters, and the like employ high resistant elements on purpose; that is, the higher the resistance, the higher the wattage, and high wattage means high energy output and heat! Conversely, electrical wiring should be of low resistance for reasons of safety and cost — hence, the use of copper, one of the best electrical conductors known. We measure resistance in *ohms*, after the German physicist Georg Simon Ohm (1787-1854), and employ the Greek symbol omega (Ω) in interesting contrast to the volt (V), ampere (A), watt (W), and the others. By SI definition, one ohm is the electric resistance between two points of a conductor when a constant difference of potential of one volt, applied between these two points, produces in this conductor a current of one ampere, or $\Omega = V/A$. Further, by switching V and A we come up with the siemens ($S = A/V$), the SI unit of electric *conductance* named for Ernst Werner von Siemens (1816-1892). As stressed by the formula, resistance and conductance are mathematical reciprocals, high resistance meaning *low* conductance and low resistance meaning *high* conductance.

11

RADIAN AND STERADIAN

Suppose we have a line, say two inches long. Suppose, further, this line splits into two lines, one of which rotates (in the same *plane*) about the end point (the vertex) for one complete revolution. In the course of this rotation the angle between the two lines gets larger and larger and reaches its acme at the end of the revolution. Not only does this maneuver generate an ever increasing plane angle but also describes a circle, a circle like any and all circles containing 360°.

And so there we have the degree, or sexagesimal, system of angular measure, perhaps man's most significant single metrological discovery. At the very least, it was the first measure of sophistication, and all agree that the full credit goes to the Sumerians of the misty past (circa 5000 *B.C.*). With this measure, for example, the Greek mathematician and astronomer Erastosthenes (third century *B.C.*) accurately estimated the circumference of the earth. Again, when the headings for the 5,000-foot Musconetcong tunnel were driven from opposite sides of the mountain the error in alignment was found to be only one half inch. Notwithstanding these marvels of triangulation, however, modern mathematics and its applications call for the *radian* (*rad*) and *steradian* (*sr*), both of which have now been adopted (1960) as supplementary SI units. The radian and steradian are certainly not new to geometry, but they are "new" in the context of international metrology. According to the American Society for Testing and Materials, "The use of the arc degree and its decimal submultiples is permissible when the radian is *not* a convenient unit. Solid angles *should* be expressed in steradians." (My italics.)

To appreciate these supplementary units we must consider *pi* (π), the transcendental number, approximately 3.14159, representing the ratio of the circumference to the diameter of a circle (and, to boot, appearing as a constant in a wide range of mathematical problems). In the radian system of angular measure, each revolution about a point is subdivided into 2π equal angles called *radians*; in other words, $360° = 2\pi$ radians. Since the circumference of a circle is equal to the *radius* times 2π, it follows that a central angle (i.e., one with its vertex at the center of a circle) is *one radian* if it is subtended by an arc (or segment of the circumference) whose length is *equal* to the radius of the circle. (In terms of degrees, the radian is equal to $360/2\pi$, or *approximately* $57°17'44.6''$*.) SI derived units of the radian include, by way of example, *radian per second (rad/s)* for angular velocity, and *radian per second per second (rad/s^2)* for angular acceleration.

Whereas the radian deals with plane angles (or simply "angles"), the steradian, as the term indicates, treats solid angles. And in the framework of what has been said about

* 57 degrees, 17 *minutes*, and 44.6 *seconds*.

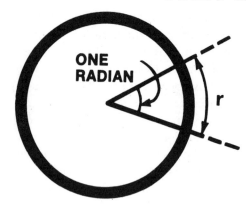

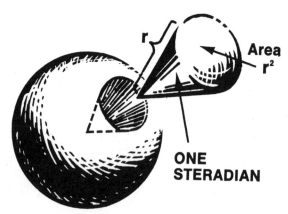

The radian (top illustration), a supplementary unit, is a plan angle
with its apex at the center of a circle and subtended by an arc equal in
length to the radius. The steradian (bottom illustration), a supplementary unit, is a solid angle with its vertex at the center of a sphere and
enclosing an area of the spherical surface equal to that of a square with
sides equal in length to the radius.

the radian the SI definition is readily understood: The steradian is the solid angle with its vertex at the center of a *sphere* that subtends an area on the spherical surface equal to that of a square with sides equal in length to the *radius*. SI derived units include the *lumen* (lm), the luminous flux emitted in a solid angle of *one* steradian by a point source having a uniform intensity of *one* candela (i.e., *lm = cd·sr*), and the *watt per steradian (W/sr)* for radiant intensity.

Part Three: TABLES AND CONVERSIONS

12

U . S .   C U S T O M A R Y S Y S T E M

LENGTH

UNIT	SYMBOL	VALUE
inch	in	——
foot	ft	12 inches
yard	yd	3 feet
rod	——	16½ feet
mile (land)	mi	5,280 feet
mile (nautical)	mi	1.151 land miles

AREA

UNIT	SYMBOL	VALUE
square inch	in^2	——
square foot	ft^2	144 square inches
square yard	yd^2	9 square feet
acre	——	43,560 square feet
square mile	mi^2	640 acres

89

VOLUME

UNIT	SYMBOL	VALUE
cubic inch	in^3	——
cubic foot	ft^3	1,728 cubic inches
cubic yard	yr^3	27 cubic feet

LIQUID MEASURE

UNIT	SYMBOL	VALUE
minim	m	——
dram	dr	60 minims
fluid ounce	oz	8 fluid drams
pint	pt	16 fluid ounces
quart	qt	2 pints
gallon	gal	4 quarts
barrel	bar	31 or 42 gallons

DRY MEASURE

UNIT	SYMBOL	VALUE
pint	pt	——
quart	qt	2 pints
peck	pk	8 quarts
bushel	bu	4 pecks

WEIGHT (AVOIRDUPOIS)

UNIT	SYMBOL	VALUE
grain	gr	——
dram	dr	27.344 grains
ounce	oz	16 drams
pound	lb	16 ounces (7,000 grains)
ton (short)	t	2000 pounds
ton (long)	t	2240 pounds

13

APOTHECARY
SYSTEM*

WEIGHT

UNIT	SYMBOL	VALUE
scruple	sc	20 grains
dram	dr	3 scruples
ounce	oz	8 drams
pound	lb	12 ounces

PHARMACEUTICAL SYMBOLS**

WEIGHT	SYMBOL	MEASURES	SYMBOL
grain	gr	minim	M
scruple	Э	fluid dram	fȝ
dram	ȝ	fluid ounce	fȝ
ounce	ȝ	pint	O
		gallon	C

* Liquid measure same as U.S. Customary.
** Used on prescriptions during the days of the apothecary system.

14

SYSTÈME INTERNATIONAL D'UNITÉS (SI)

BASE UNITS*

QUANTITY	UNIT	SYMBOL
length	meter	m
mass	kilogram	kg
time	second	s
electric current	ampere	A
temperature	kelvin	K
amount of substance	mole	mol
luminous intensity	candela	cd

SUPPLEMENTARY UNITS

plane angle	radian	rad
solid angle	steradian	sr

* Unit names are not capitalized, not even the proper names *ampere* and *kelvin*. Also, the plural is used; e.g., "100 kelvins". Symbols are not capitalized, except for ampere (A) and kelvin (K), and no periods.

DERIVED UNITS**

QUANTITY	UNIT	SYMBOL
acceleration	meter per second/s	——
area	square meter	——
density	kilogram per cubic meter	——
electric resistance	ohm	Ω
energy	joule	J
force	newton	N
frequency	hertz	Hz
illuminance	lux	lx
luminance	candela per square meter	——
luminous flux	lumen	lm
power	watt	W
pressure	pascal	Pa
quantity of heat	joule	J
stress	pascal	Pa
velocity	meter per second	——
voltage	volt	V
volume	cubic meter	——
work	joule	J

** The more commonly encountered units are listed, including most of those with special names (which carry a *symbol*). Derived units without special names are formed as needed from base units or other derived units or both.

15

APPROXIMATE CONVERSIONS FROM U. S. CUSTOMARY TO SI AND VICE VERSA

LENGTH

	Multiply by
inches to millimeters	25*
feet to centimeters	30
yards to meters	0.9
miles to kilometers	1.6
millimeters to inches	0.04
centimeters to inches	0.4
meters to yards	1.1
kilometers to miles	0.6

* For example, 3 inches equals 75 millimeters.

AREA

	Multiply by
square inches to square centimeters	6.5
square feet to square meters	0.09
square yards to square meters	0.8
square miles to square kilometers	2.6
acres to square hectometers (hectares)	0.4
square centimeters to square inches	0.16
square meters to square yards	1.2
square kilometers to square miles	0.4
square hectometers (hectares) to acres	2.5

WEIGHT (MASS)

	Multiply by
ounces to grams	28.3
pounds to kilograms	0.425
short tons to megagrams (metric tons)	0.9
grams to ounces (avoir.)	0.035
kilograms to pounds	2.2
megagrams (metric tons) to short tons	1.1

LIQUID MEASURE

	Multiply by
ounces to milliliters	30
pints to liters	0.47
quarts to liters	0.95
gallons to liters	3.8
milliliters to ounces	0.034
liters to pints	2.1
liters to quarts	1.06
liters to gallons	0.26

16

COMMON CONVERSIONS ACCURATE TO SIX SIGNIFICANT FIGURES

LENGTH

	Multiply by
inches to millimeters	25.4
feet to meters	0.3048
yards to meters	0.9144
miles to kilometers	1.60934
millimeters to inches	0.039370
meters to feet	3.28084
meters to yards	1.09361
kilometers to miles	0.621371

AREA

	Multiply by
square yards to square meters	0.836127
acres to hectares*	0.404686
square meters to square yards	1.19599
hectares to acres	2.47106

WEIGHT

	Multiply by
ounces (avoir.) to grams	28.3495
pounds (avoir.) to kilograms	0.453592
grams to ounces (avoir.)	0.035274
kilograms to pounds (avoir.)	2.20462

VOLUME

	Multiply by
cubic yards to cubic meters	0.764555
quarts (lq) to liters	0.946353
cubic meters to cubic yards	1.30795
liters to quarts (lq)	1.05669

* Hectares is the common name for 10,000 square meters.

17

TEMPERATURE CONVERSION "BY FRACTION"

I. Degrees *Fahrenheit* ($^\circ$F) to degrees *Celsius* ($^\circ$C):
$$^\circ C = 5/9 \ (^\circ F - 32)*$$
Example: 212°F (boiling) on the Celsius scale is:
$$C = 5/9 \ (212 - 32)$$
$$C = 5/9 \ (180)$$
$$C = \frac{900}{9}$$
$$C = 100^\circ$$

II. Degrees *Celsius* ($^\circ$C) to degrees *Fahrenheit* ($^\circ$F):
$$F = 9/5 \ ^\circ C + 32*$$
Example 0° C (freezing) on the Fahrenheit scale is:
$$F = 9/5 \ (0^\circ) + 32$$
$$F = 0 + 32$$
$$F = 32^\circ$$

* 1 *degree* on the Celsius scale equals 1.8° (9/5) degrees on the Fahrenheit scale; or, turned around, 1°F equals 5/9 $^\circ$C. To convert a given *temperature* from one scale to the other, however, we must also take into account the difference in *freezing points* — hence the figure 32.

18

TEMPERATURE CONVERSION "BY DECIMAL"

I. Degrees *Fahrenheit* (°F) to degrees *Celsius* (°C):

$$°C = \frac{°F-32*}{1.8}$$

Example: 212° F (boiling) on the Celsius scale is:

$$C = \frac{212-32}{1.8}$$

$$C = \frac{180}{1.8}$$
$$C = 100°$$

II. Degrees *Celsius* (°C) to degrees *Fahrenheit* (°F):

$$°F = 1.8°C + 32*$$

Example: 0°C (freezing) on the Fahrenheit scale is:

$$F = 1.8(0) + 32$$
$$F = 32°$$

---

* 1 *degree* on the Celsius scale equals 1.8° (9/5) degrees on the Fahrenheit scale; or, turned around, 1°F equals 5/9 °C. To convert a given *temperature* from one scale to the other, however, we must also take into account the difference in *freezing points* — hence the figure 32.

19

COMPARISON OF CALCULATIONS— CUSTOMARY VS. METRIC*

EXAMPLE: CARPETING

Customary units:

Calculate the amount of carpeting to buy for wall-to-wall carpeting of a room 18 feet 4 inches long and 11 feet 8 inches wide, using carpet 12 feet wide.

$Area = length \times width$

$$= \left(\frac{18}{3} + \frac{4}{36}\right) \times \left(\frac{12}{3}\right)$$

$= 24.44$ *square yards to buy*

Metric units:

Calculate the amount of carpeting to buy for wall-to-wall carpeting of a room 5.59 meters long and 3.81 meters wide, using carpet 4 meters wide.

> *Area = length* x *width*
>   = 5.59 x 4
>   = *22.36 square meters to buy*

* Courtesy National Bureau of Standards

GLOSSARY

ABSOLUTE ZERO — the temperature (0 on absolute scale) at which molecules possess minimal energy (equal to $-273.15°$C or $-459.67°$F).

ACCELERATION — change of speed; determined by dividing change by time it takes for change to occur.

ACRE — 43,560 square feet.

AMPERE (A) — SI unit of electric current. Defined as that constant current which, if maintained in two straight parallel conductors of infinite length, of negligible cross section, and placed one meter apart in a vacuum, would produce between these conductors a magnetic force equal to 2×10^{-7} newtons per meter of length.

AMPHORA — Roman unit of liquid measure equal to about six gallons.

ANGLE — figure formed by two lines arising from a common point.

APOTHECARIES' SYSTEM — system of weights and measures used in pharmacy; measures same as U.S. Customary, but weights are based on an ounce equal to 480 grains and a pound equal to 12 ounces (5,760 grains).

103

ATOMIC CLOCK — timepiece regulated by an invariant frequency of an atomic or molecular system.

ATTO- (*a*) — SI prefix denoting 10^{-18}.

AVOGADRO'S NUMBER — the number of molecules in a mole of a substance, approximately equal to 6.023×10^{23}.

AVOIRDUPOIS WEIGHT — the system of weights (of U.S. Customary System) based on an ounce equal to 437.59 grains and a pound equal to 16 ounces (7,000 grains).

BARREL — in the U.S. Customary System a unit of liquid measure varying from 31 to 42 gallons.

BLACKBODY — perfect absorber of all incident radiation.

BRITISH IMPERIAL SYSTEM — one of the three major systems of measurement; like the U.S. Customary System the fundamental units are the yard and pound.

BRITISH THERMAL UNIT (*Btu*) — amount of heat needed to raise the temperature of one pound of water $1°F$.

BROWNIAN MOTION (MOVEMENT) — rapid oscillatory motion often observed in very minute particles suspended in water, air or other fluid.

BUSHEL (*bu*) — unit of dry measure in U.S. Customary System equal to 4 pecks (2,150.42 cubic inches); unit of dry and liquid in British Imperial System (2,219.36 cubic inches).

CANDELA (*cd*) — SI unit of luminous intensity. Defined as the intensity of 1/600,000 of a square meter of a blackbody at the temperature of freezing platinum under a pressure of 101,325 newtons per square meter.

CALORIE — small calorie (*cal*) is amount of heat needed to raise the temperature of one gram of water $1°$Celsius;

large calorie *(Cal)* is amount of heat needed to raise the temperature of one kilogram of water 1°Celsius.

CARAT — unit of weight for precious stones equal to 200 milligrams.

CELSIUS *(C)* — temperature scale on which water freezes at 0°C and boils at 100°C under normal atmospheric pressure.

CENTI - *(c)* — SI prefix denoting 10^{-2}.

CENTIGRADE *(C)* — the Celsius temperature scale.

CONCENTRATION — amount of a substance in a unit amount of another substance.

CONGIUS *(C)* — Roman unit of liquid measure equal to about three quarts.

CUBIC METER — SI unit of volume.

CUBIT — Egyptian hieroglyphic equal to the length of the forearm from the tip of the middle finger to the elbow.

CYCLE — time interval in which an event or sequence of events occurs.

DENSITY — mass per unit volume.

DERIVED UNIT — unit of measure based on a fundamental unit.

DECI- *(d)* — SI prefix denoting 10^{-1}.

DEKA- *(da)* — SI prefix denoting 10^{1}.

DRAM *(dr)* — 1/8 fluid ounce; 1/8 ounce apothecary; 1/16 ounce avoirdupois.

ELECTROMAGNETIC SPECTRUM — range of radiation, including, in order of decreasing frequency (and increasing wavelength), gamma rays, x rays, ultraviolet radiation, visible light, infrared radiation, microwaves, radio waves, and heat.

ENERGY — the ability to do work.

ENGLISH SYSTEM — the general term encompassing the British Imperial System and U.S. Customary System.

FATHOM — originally the distance between the tips of the middle fingers in the outstretched arms. Today, a unit of length equal to six feet.

FAHRENHEIT (F) — temperature scale on which water freezes at $32°F$ and boils at $212°F$ under normal atmospheric pressure.

FEMTO- (f) — SI prefix denoting 10^{-15}.

FOOTCANDLE — illumination of a surface one foot distant from a source of one candela.

FORCE — directional quantity that tends to produce acceleration.

FREQUENCY — number of times a specified event occurs within a specified interval.

FURLONG — 220 yards.

GALLON (*gal*) — unit of volume or capacity in the U.S. Customary System, used in liquid measure, equal to 4 quarts (231 cubic inches); unit of volume in British Imperial System, used in liquid and dry measure, equal to 277.42 cubic inches.

GIGA- (G) — SI prefix denoting 10^9.

GRAIN (*gr*) — unit of weight in U.S. Customary System equal to 0.036 dram; 7,000 grains equal an avoirdupois pound, and 5,760 grains equal an apothecary pound.

GRAM (*g*) — metric unit of mass and weight equal to 0.001 kilogram.

GRAM MOLECULAR WEIGHT — the molecular weight of a substance expressed in grams; same as mole.

HECTO- (*h*) — SI prefix denoting 10^2.

HERTZ (*Hz*) — SI unit of frequency equal to one cycle per second.

HOGSHEAD (*hhd*) — unit of capacity used in liquid measure in the United States equal to 63 gallons.

HORSEPOWER (*hp*) — 33,000 foot-pounds of work per minute.

INTERFEROMETER — instrument for measuring small lengths or distances by means of the interference of two rays of light.

INGERUM — 28,800 square feet (ancient Rome).

ISOTOPES — atoms which are alike except in regard to their weights.

JOULE (*J*) — SI fundamental unit of energy, work and quantity of heat; joules equal newtons times meters.

KELVIN (*K*) — SI unit of temperature. Defined as the fraction 1/273.16 of the thermodynamic temperature of the triple point of water.

KILO- (*k*) — SI prefix denoting 10^3.

KILOGRAM (*kg*) — SI fundamental unit of mass and weight equal to 2.20462 pounds.

KRYPTON (*kr*) — rare and inert gaseous element of atmosphere.

LIBRA — unit of weight in ancient Rome corresponding to a pound and equivalent to approximately 12 ounces.

LITER — a metric unit of fluid measure (liquid and gas) equal to 0.001 cubic meter.

LUMEN (*lm*) — SI unit of light (luminous) flux. One lumen is the luminous flux emitted in a solid angle of one steradian by a point source having an intensity of one candela.

LUX (*lx*) — SI unit of illumination produced by a luminous flux of one lumen uniformly distributed over a surface of one square meter.

MASS — quantity of matter.

MEAN SOLAR DAY — interval between two successive passages of the mean (hypothetical) solar sun.

MEGA- (*M*) — SI prefix denoting 10^6.

MERIDIAN — great circle on the earth's surface passing through both geophysical poles.

METER (*m*) — SI fundamental unit of length equal to 1,650,763.73 wavelengths in a vacuum of the orangered radiation of krypton 86.

METRICATION — any act tending to increase the use of the metric system (SI).

METRICIZE — to convert any other unit to its metric (SI) equivalent.

METRIC SYSTEM — decimal system of weights and measures based on the meter as unit of length and the kilogram as unit of mass; now used synonymously with SI.

METRIC TON — 1 megagram.

MICRO- (*μ*) — SI prefix denoting 10^{-6}.

MICROWAVES — electromagnetic radiations with wavelengths

in the region between infrared and short-wave radio wavelengths.

MILE (*mi*) — unit of length equal to 5,280 feet; also called land mile or statute mile.

MILLI- (*m*) — SI prefix denoting 10^{-3}.

MINA —the earliest known basic unit of weight (Mesopotamia)

MINIM (*M*) — 1/60 fluid dram.

MODIUS — unit of dry measure equal to about a peck (ancient Rome).

MOLE (*mol*) — SI fundamental unit for amount of substance; one mole is amount of substance of a system that contains as many elementary entities as there are atoms in 0.012 kilogram of carbon 12.

MOLECULE — a combination of two or more atoms; the smallest particle of a compound.

MOLECULAR WEIGHT — the sum of the atomic weights of a molecule's constituent atoms.

MONOCHROMATIC — producing light of only one wavelength.

NANO- (*n*) — SI prefix denoting 10^{-9}.

NAUTICAL MILE (*nm*) — unit of length used in sea and air navitagion (based on length of one minute of arc of a great circle) equal to 1.151 statute miles; also called international mile.

NEWTON (*N*) — SI unit of force; newtons equal kilograms times meters divided by seconds squared.

OHM (Ω) — SI unit of electrical resistance. Defined as resistance between two points of a conductor when a constant difference of potential of one volt, applied between these two points, produces in this conductor a current of one ampere.

OUNCE (*oz*) — an avoirdupois unit (U.S. Customary System) equal to 16 drams or 437.5 grains; an apothecary unit equal to 480 grains; a unit of liquid measure (U.S. Customary System) equal to 8 fluid drams.

PASCAL (*Pa*) — SI unit of pressure; pascals equal newtons divided by square meters.

PECK (*pk*) — unit of dry measure in U.S. Customary System equal to 8 quarts (537.605 cubic inches); unit of dry and liquid measure in British Imperial System equal to 554.84 cubic inches.

PI (*π*) — transcendental number, approximately 3.14159, representing the ratio of the circumference to the diameter of a circle and appearing as a constant in a wide range of mathematical problems.

PICO- (*p*) — SI prefix denoting 10^{-12}.

PINT (*pt*) — 1/2 quart.

PLANE — surface containing all the straight lines connecting any two points on it.

POUND (*lb*) — fundamental unit of weight in U.S. Customary System and British Imperial System. Avoirdupois (official) pound is equal to 16 ounces (7,000 grains); apothecary (troy) pound is equal to 12 ounces (5,760 grains).

POWER — rate at which work is done.

PRESSURE — force applied over a surface.

PROTOTYPE — "an original" type that serves as a model.

QUART (*q*) — unit of liquid measure in the U.S. Customary System equal to two pints (57.75 cubic inches); unit of dry measure in U.S. Customary System equal to two pints (or 67.2 cubic inches).

RADIAN *(rd)* — plane angle with its vertex at the center of a circle that is subtended by an arc equal in length to the radius.

RESONATE — the enhancement of the response of a physical system when the driving frequency is equal to the natural frequency of the system.

SCRUPLE *(sc)* — unit of apothecary weight equal to 20 grains.

SECOND *(s)* — SI fundamental unit of time defined as the duration of 9,192,631,770 cycles of the radiation associated with a specified transition of the cesium −133 atom.

SEXAGESIMAL — relating to or based upon the number 60.

SEXTARIUS — unit of liquid measure equal to about a pint (ancient Rome).

SHEKEL — a subdivision of a mina.

SI — official abbreviation for Système International d'Unités.

SIEMENS *(S)* — SI unit of electric conductance equal to amperes divided by volts.

SOLAR DAY— interval between two meridian passages of the sun.

SOLID ANGLE — angle formed by three or more planes intersecting in a common point or at the vertex of a cone.

SPEED — distance traveled divided by the time of travel.

SPHERE — three-dimensional surface, all points of which are equidistant from a fixed point.

SQUARE METER — SI unit of area.

STANDARD OF MEASUREMENT — an object or natural physical phenomenon which serves to define the magnitude of a unit of measurement.

STERADIAN (*sr*) — solid angle with its vertex at the center of a sphere that is subtended by an area of the spherical surface equal to that of a square with sides equal to the radius.

TALENT — a variable unit of weight and money used in ancient Greece, Rome, and Middle East.

TERA- (*T*) — SI prefix denoting 10^{12}.

TON — unit of weight in U.S. Customary System equal to either 2,240 pounds ("long ton") or 2,000 pounds ("short ton").

TRIPLE POINT — coexistence of three states of matter.

TROY WEIGHT — same as apothecary weight.

UNICA — "twelfth part," in Latin: forerunner of the inch and ounce.

UNIT — precisely specified quantity in terms of which the magnitude of other quantities of the same kind are stated.

U.S. CUSTOMARY SYSTEM — the "yard-pound" system of weights and measures used in the United States.

VELOCITY — speed in a given direction.

VERTEX — point at which the sides of an angle intersect.

VOLT (*V*) — SI unit of voltage or electromotive force. Defined as potential difference between two points on a conducting wire carrying a constant current of one ampere when the power dissipated between the two points is one watt.

WATT (*W*) — SI unit of power; watts equal joules divided by seconds.

WAVELENGTH — distance between two points of corresponding phase in consecutive cycles.

WEIGHT — variable force equal to mass times gravitational acceleration.

WORK — transfer of energy from one body to another; quantity of work is the product of the force acting upon a body and the distance through which the point of application of force moves.

YARD (*yd*) — the fundamental unit of length in the U.S. Customary System and British Imperial System equal to 0.9144 meter.

BIBLIOGRAPHY

Barbrow, L. E.: *What about metric*, National Bureau of Standards (U.S.) Conference on Information (1073).

Blair, B. E.: *Time and frequency: theory and fundamentals*, National Bureau of Standards (U.S.) Monograph 140 (January 1974).

Brief history of measurement systems, National Bureau of Standards (U.S.) Special Publication 304A, 1968 (Revised September 1974).

Brixley, J. C. and Andree, R. V.: *Fundamentals of college mathematics*, edition 2, New York, 1961, Holt, Rinehart and Winston.

Burton, W. K., editor: *Measuring systems and standards organizations*, New York, American Standards Institute.

"Changing to the metric system, conversion factors, symbols, definitions," National Physical Laboratory, London, 1965, Her Majesty's Stationery Office.

"Conversion factors and tables," British Standard 350, Part 1 (1959).

"Conversion factors and tables," British Standard 350, Part 2 (1962).

De Simone, D. V.: *A metric America: a decision whose time has come*, National Bureau of Standards (U.S.) Special Publication 345 (July 1971).

For good measure: metric and customary units, National Bureau of Standards (U.S.) Special Publication 376 (1972).

Fundamental physical constants, National Bureau of Standards (U.S.) Special Publication 398 (1974).

Gerber, E. A., and Sykes, R. A.: *A quarter century of progress in the theory of development of crystals for frequency control and selection*, Proceedings of 25th Annual Symposium on Frequency Control, Ft. Monmouth, N.J. (1971).

Going metric: the first 5 years 1965-1969, the first report of the metrication board 1970, London, 1970, Her Majesty's Stationery Office.

Gordon, G.: *International trade,* National Bureau of Standards (U.S.) Special Publication 345-8 (1971).

Hatos, S.: *Commercial weights and measures,* National Bureau of Standards (U.S.) Special Publication 345-3 (1971).

Hellwig, H.: *Frequency standards and clocks: a tutorial introduction,* National Bureau of Standards (U.S.) Technical Note 616 (Revised March 1974).

Huntoon, R., et al: *International standards,* National Bureau of Standards (U.S.) Special Publication 345-1 (1970).

Judson, L. V.: *Weights and measures standards of the United States,* a brief history, National Bureau of Standards (U.S.) Miscellaneous Publication 247 (October 1963).

McNish, A.G.: *The international system of units,* Materials Research and Standards, American Society for Testing and Materials, Volume 5, Number 10 (October 1965).

Metric changeover, the, National Bureau of Standards (U.S.) Technical News Bulletin, Volume 57, Number 5 (May 1973).

Metric conversion factors, National Bureau of Standards (U.S.) Letter Circular 1051 (July 1973).

Metric conversion cards, National Bureau of Standards (U.S.) Special Publication 365 (Revised, November 1972).

Metric practice guide E 380-72, Philadelphia, 1972, American Society for Testing and Materials.

Metric standards for engineering, British Standard Handbook No. 18 (1966).

Modernized metric system, the, (chart), National Bureau of Standards (U.S.) Publication 304 (Revised October 1972).

Morris, W., editor: *The American Heritage Dictionary of the English Language,* Boston and New York, 1969, American Heritage Publishing Co., Inc.

Orientation for company metric studies, edition 2, New York, 1970, American National Standards Institute.

Page, C. H., and Vigoureux, P.: *The international system of units* (SI), National Bureau of Standards (U.S.) Publication 330 (1972).

"Policy for National Bureau of Standards usage of SI units," National Bureau of Standards (U.S.) News Bulletin, Volume 55, Number 1, (January 1971).

Robinson, B.: Education, National Bureau of Standards (U.S.) Special Publication 345-6 (1971).

Treat, C.: *A history of the metric system controversy in the United States,* National Bureau of Standards (U.S.) Special Publication 345-10 (1971).

Units and systems of weights and measures. Their origin, development and present status, National Bureau of Standards (U.S.) Circular LC1035 (January 1960).

Units *of weights and measures* (U.S. Customary and metric): definitions and tables of equivalents, National Bureau of Standards (U.S.) Miscellaneous Publication MP233 (December 1960).

Viezbicke, P. P.: *National Bureau of Standards and time broadcast services,* National Bureau of Standards (U.S.) Special Publication 236 (1971).

Wall chart of the modernized metric system, National Bureau of Standards (U.S.) Special Publication 304 (1972).

What about metric?, National Bureau of Standards (U.S.) CIS 7 (October 1973).

INDEX

119